GRADE 5 MATH WORKBOOK

with ANSWERS

decimal place values

decimal arithmetic

fraction arithmetic

geometric figures

financial math

$$\begin{array}{r} 2.475 \\ \times\ 0.3 \\ \hline 0.7425 \end{array}$$

multiplication/division

pattern recognition

data analysis

scatter plots

unit conversions

9, 16, 23, 30, 37, 44, __, __

Chris McMullen, Ph.D.

Grade 5 Math Workbook with Answers

Chris McMullen, Ph.D.

www.monkeyphysicsblog.wordpress.com

www.improveyourmathfluency.com

www.chrismcmullen.wordpress.com

Zishka Publishing

ISBN: 978-1-941691-33-5

Mathematics > Arithmetic

Study Guides > Workbooks > Math

Education > Math > Grade 5

CONTENTS

INTRODUCTION

This workbook covers a variety of topics that are typically taught in fifth grade mathematics. Each section begins by reviewing essential terms, concepts, and problem-solving strategies. The techniques are illustrated in examples that should serve as a helpful guide for completing the exercises at the end of the section. The answer to every problem can be found in the Answer Key at the back of the book. Topics include:

- decimal place value
- multiplying and dividing decimals
- arithmetic with fractions
- interpreting data and graphs
- number pattern recognition
- interpreting histograms
- measurements and geometric figures
- prime numbers versus composite numbers
- and much more

-1-

MULTIPLICATION

1.1 Quick review of multiplication facts

Fluency with arithmetic facts is essential in math.

Problems. Practice these multiplication facts.

① $4 \times 6 =$	② $7 \times 5 =$	③ $8 \times 9 =$
④ $7 \times 7 =$	⑤ $9 \times 6 =$	⑥ $9 \times 4 =$
⑦ $5 \times 8 =$	⑧ $7 \times 4 =$	⑨ $6 \times 5 =$
⑩ $9 \times 7 =$	⑪ $8 \times 8 =$	⑫ $5 \times 7 =$
⑬ $6 \times 6 =$	⑭ $4 \times 9 =$	⑮ $6 \times 8 =$
⑯ $9 \times 5 =$	⑰ $7 \times 6 =$	⑱ $4 \times 5 =$
⑲ $8 \times 4 =$	⑳ $5 \times 5 =$	㉑ $9 \times 9 =$
㉒ $6 \times 7 =$	㉓ $9 \times 8 =$	㉔ $7 \times 8 =$
㉕ $4 \times 4 =$	㉖ $7 \times 9 =$	㉗ $9 \times 3 =$
㉘ $8 \times 6 =$	㉙ $3 \times 8 =$	㉚ $6 \times 9 =$

1.2 Two digits times one digit

Follow these steps to multiply a two-digit number by a one-digit number:

$$\begin{array}{r} 36 \\ \times 4 \\ \hline \end{array}$$

First multiply $6 \times 4 = 24$.

$$\begin{array}{r} {}^{2} \\ 36 \\ \times 4 \\ \hline 4 \end{array}$$

Write the 4 of the 24 on the bottom right.
Write the 2 of the 24 above the tens digit of the top number.

$$\begin{array}{r} {}^{2} \\ 36 \\ \times 4 \\ \hline 4 \end{array}$$

Now multiply $3 \times 4 = 12$.
Add the 2 from above: $12 + 2 = 14$.

$$\begin{array}{r} {}^{2} \\ 36 \\ \times 4 \\ \hline 144 \end{array}$$

Write the 14 to the left of the bottom number.

The final answer is 144.

Here is a different way of looking at it:

- 36×4 is the same as $(30 + 6) \times 4$.
- $30 \times 4 = 120$ and $6 \times 4 = 24$.
- Add these together: $120 + 24 = 144$.

This method applies the distributive property (Sec. 1.13). The idea is that 36 has 3 tens and 6 units. When we multiply 36 by 4, we get 12 tens (since $3 \times 4 = 12$) and 24 units (since $6 \times 4 = 24$). Note that 12 tens makes 120 (since $12 \times 10 = 120$), such that 12 tens plus 24 units makes 144 units.

Examples. Multiply the numbers.

(A)

$$\begin{array}{r} {}^{1} \\ 53 \\ \underline{\times\, 6} \\ 318 \end{array}$$

$3 \times 6 = 18$ Write the 1 above and the 8 below.
$5 \times 6 = 30$ Remember to add the 1 above.
$30 + 1 = 31$ Write the 31 on the bottom left.

(B)

$$\begin{array}{r} 43 \\ \underline{\times\, 2} \\ 86 \end{array}$$

In this example, the first answer only has 1 digit.
$3 \times 2 = 6$ Write the 6 below. (Nothing goes above.)
$4 \times 2 = 8$ Write the 8 on the bottom left.
(There is nothing to add in this problem.)

(C)

$$\begin{array}{r} {}^{4} \\ 26 \\ \underline{\times\, 7} \\ 182 \end{array}$$

$6 \times 7 = 42$ Write the 4 above and the 2 below.
$2 \times 7 = 14$ Remember to add the 4 above.
$14 + 4 = 18$ Write the 18 on the bottom left.

Problems. Multiply the numbers.

① 47×3

② 29×5

③ 54×6

④ 12×4

⑤ 31×8

⑥ 75×7

⑦ 68×9

⑧ 83×2

⑨ 46×7

⑩ 97×8

1.3 Three digits times one digit

Follow these steps to multiply a three-digit number by a one-digit number:

$$\begin{array}{r} {\scriptstyle 1} \\ 495 \\ \times 3 \\ \hline 5 \end{array}$$

First multiply $5 \times 3 = 15$.
Write the 5 of the 15 on the bottom right.
Write the 1 of the 15 above the tens digit of the top number.

$$\begin{array}{r} {\scriptstyle 2\,1} \\ 495 \\ \times 3 \\ \hline 85 \end{array}$$

Now multiply and add $9 \times 3 + 1 = 27 + 1 = 28$.
Write the 8 of the 28 on the bottom.
Write the 2 of the 28 above the hundreds digit of the top number.

$$\begin{array}{r} {\scriptstyle 2\,1} \\ 495 \\ \times 3 \\ \hline 1485 \end{array}$$

Now multiply and add $4 \times 3 + 2 = 12 + 2 = 14$.
Write the 14 on the bottom left.

The final answer is 1485.

Here is a different way of looking at it:

- 495×3 is the same as $(400 + 90 + 5) \times 3$.
- $400 \times 3 = 1200$, $90 \times 3 = 270$, and $5 \times 3 = 15$.
- Add these together: $1200 + 270 + 15 = 1485$.

Example. Multiply the numbers.

(A)
$$\begin{array}{r} {}^{1\,2} \\ 237 \\ \times\ 4 \\ \hline 948 \end{array}$$

$7 \times 4 = 28$ Write the 2 above and the 8 below.

$3 \times 4 + 2 = 12 + 2 = 14$ Write the 1 above and the 4 below.

$2 \times 4 + 1 = 8 + 1 = 9$ Write the 9 on the bottom.

Problems. Multiply the numbers.

①
$$\begin{array}{r} 263 \\ \times\ 5 \\ \hline \end{array}$$

②
$$\begin{array}{r} 459 \\ \times\ 2 \\ \hline \end{array}$$

③
$$\begin{array}{r} 617 \\ \times\ 7 \\ \hline \end{array}$$

④
$$\begin{array}{r} 384 \\ \times\ 8 \\ \hline \end{array}$$

⑤
$$\begin{array}{r} 928 \\ \times\ 6 \\ \hline \end{array}$$

⑥
$$\begin{array}{r} 875 \\ \times\ 3 \\ \hline \end{array}$$

⑦
$$\begin{array}{r} 792 \\ \times\ 4 \\ \hline \end{array}$$

⑧
$$\begin{array}{r} 536 \\ \times\ 9 \\ \hline \end{array}$$

1.4 Two digits times two digits

Follow these steps to multiply a two-digit number by a two-digit number:

```
   1
  43
× 75
 215
```

First multiply 43×5.
$3 \times 5 = 15$ Write the 1 above and the 5 below.
$4 \times 5 + 1 = 20 + 1 = 21$ Write the 21 below.

```
   1
  43
× 75
 215
   0
```

Next write a 0 on the bottom right.
Why? Because next we will multiply by the tens digit. When you multiply by a ten, the answer always ends with zero.

```
   2
   1
  43
× 75
 215
3010
```

Now multiply 43×7.
$3 \times 7 = 21$ Write the 2 above and the 1 below.
$4 \times 7 + 2 = 28 + 2 = 30$ Write the 30 below.
We added the 2 to 28. (The 1 was used earlier.)

```
   2
   1
  43
× 75
 215
3010
3225
```

Finally, add 215 and 3010.
$215 + 3010 = 3225$ Write this at the bottom.

Here is another way of looking at it:

- 43×75 is the same as $43 \times (70 + 5)$.
- $43 \times 70 = 3010$ and $43 \times 5 = 215$. (Note that 43×70 is ten times as large as $43 \times 7 = 301$.)
- Add these together: $3010 + 215 = 3225$.

Examples. Multiply the numbers.

(A)

$$\begin{array}{r} {}^{2}_{1} \\ 56 \\ \times\ 43 \\ \hline 168 \\ 2240 \\ \hline 2408 \end{array}$$

$6 \times 3 = 18$ Write the 1 above and the 8 below.
$5 \times 3 + 1 = 15 + 1 = 16$ Write the 16 below.
Write a 0 below 168 on the right.
$6 \times 4 = 24$ Write the 2 above and the 4 below.
$5 \times 4 + 2 = 20 + 2 = 22$ Write the 22 below.
$168 + 2240 = 2408$ Write the 2408 below.

(B)

$$\begin{array}{r} {}^{1}_{2} \\ 47 \\ \times\ 23 \\ \hline 141 \\ 940 \\ \hline 1081 \end{array}$$

$7 \times 3 = 21$ Write the 2 above and the 1 below.
$4 \times 3 + 2 = 12 + 2 = 14$ Write the 14 below.
Write a 0 below 141 on the right.
$7 \times 2 = 14$ Write the 1 above and the 4 below.
$4 \times 2 + 1 = 8 + 1 = 9$ Write the 9 below.
$141 + 940 = 1081$ Write the 1081 below.

Problems. Multiply the numbers.

① $\begin{array}{r} 53 \\ \times\ 24 \\ \hline \end{array}$

② $\begin{array}{r} 37 \\ \times\ 16 \\ \hline \end{array}$

③ $\begin{array}{r} 68 \\ \times\ 45 \\ \hline \end{array}$

④ $\begin{array}{r} 92 \\ \times\ 38 \\ \hline \end{array}$

⑤ $\begin{array}{r} 36 \\ \times\ 79 \\ \hline \end{array}$

⑥ $\begin{array}{r} 81 \\ \times\ 50 \\ \hline \end{array}$

⑦

$$\begin{array}{r} 88 \\ \times\ 51 \\ \hline \end{array}$$

⑧

$$\begin{array}{r} 54 \\ \times\ 65 \\ \hline \end{array}$$

⑨

$$\begin{array}{r} 46 \\ \times\ 33 \\ \hline \end{array}$$

⑩

$$\begin{array}{r} 69 \\ \times\ 12 \\ \hline \end{array}$$

⑪

$$\begin{array}{r} 95 \\ \times\ 28 \\ \hline \end{array}$$

⑫

$$\begin{array}{r} 77 \\ \times\ 44 \\ \hline \end{array}$$

1.5 Three digits times two digits

Follow these steps to multiply a three-digit number by a two-digit number:

```
  3 2
  754
 × 36
 4524
```

$4 \times 6 = 24$ Write the 2 above and the 4 below.
$5 \times 6 + 2 = 30 + 2 = 32$ Write the 3 above and the 2 below.
$7 \times 6 + 3 = 42 + 3 = 45$ Write the 45 below.

```
  3 2
  754
 × 36
 4524
    0
```

Next write a 0 on the bottom right.
Why? Because next we will multiply by the tens digit. When you multiply by a ten, the answer always ends with zero.

```
   1 1
   3 2
   754
  × 36
 4,524
22,620
```

$4 \times 3 = 12$ Write the 1 above and the 2 below.
$5 \times 3 + 1 = 15 + 1 = 16$ Write the 1 above and the 6 below.
$7 \times 3 + 1 = 21 + 1 = 22$ Write the 22 below.
We added the 1's. (The 2 and 3 were used earlier.)

```
   1 1
   3 2
   754
  × 36
 4,524
22,620
27,144
```

Finally, add 4,524 and 22,620.
$4{,}524 + 22{,}620 = 27{,}144$ Write this at the bottom.

The final answer is 27,144.

Here is another way of looking at it:

- 754×36 is the same as $754 \times (30 + 6)$.
- $754 \times 30 = 22{,}620$ and $754 \times 6 = 4524$. (Note that 754×30 is ten times as large as $754 \times 3 = 2262$.)
- Add these together: $22{,}620 + 4{,}524 = 27{,}144$.

Example. Multiply the numbers.

(A)

$$\begin{array}{r} {\scriptstyle 2\,4} \\ {\scriptstyle 1\,3} \\ 238 \\ \times\ 64 \\ \hline 952 \\ 14{,}280 \\ \hline 15{,}232 \end{array}$$

$8 \times 4 = 32$
$3 \times 4 + 3 = 12 + 3 = 15$
$2 \times 4 + 1 = 8 + 1 = 9$
Write a 0 below 952 on the right.
$8 \times 6 = 48$
$3 \times 6 + 4 = 18 + 4 = 22$
$2 \times 6 + 2 = 12 + 2 = 14$
$952 + 14{,}280 = 15{,}232$

Problems. Multiply the numbers.

①

$$\begin{array}{r} 386 \\ \times\ 54 \\ \hline \end{array}$$

②

$$\begin{array}{r} 158 \\ \times\ 27 \\ \hline \end{array}$$

③

$$\begin{array}{r} 629 \\ \times\ 83 \\ \hline \end{array}$$

④

$$\begin{array}{r} 473 \\ \times\ 16 \\ \hline \end{array}$$

⑤

$$\begin{array}{r} 841 \\ \times\ 75 \\ \hline \end{array}$$

⑥

$$\begin{array}{r} 585 \\ \times\ 42 \\ \hline \end{array}$$

⑦

$$\begin{array}{r} 730 \\ \times\ 96 \\ \hline \end{array}$$

⑧

$$\begin{array}{r} 612 \\ \times\ 30 \\ \hline \end{array}$$

⑨

$$\begin{array}{r} 264 \\ \times\ 61 \\ \hline \end{array}$$

⑩

$$\begin{array}{r} 407 \\ \times\ 58 \\ \hline \end{array}$$

⑪

$$\begin{array}{r} 593 \\ \times\ 44 \\ \hline \end{array}$$

⑫

$$\begin{array}{r} 986 \\ \times\ 79 \\ \hline \end{array}$$

⑬

$$\begin{array}{r} 757 \\ \times\ 37 \\ \hline \end{array}$$

⑭

$$\begin{array}{r} 852 \\ \times\ 86 \\ \hline \end{array}$$

1.6 Missing digits challenge

The multiplication problems in this section are solved, but one or more digits are missing from the problem or solution. You can reason out the missing digit(s) by thinking through the steps involved in the multiplication process.

Example. Determine the missing digit.

(A)

$$\begin{array}{r} \\ 1\square4 \\ \times\,3 \\ \hline 582 \end{array} \qquad \begin{array}{r} {\scriptstyle 1}\;\, \\ 1\square4 \\ \times\,3 \\ \hline 582 \end{array} \qquad \begin{array}{r} {\scriptstyle 2\,1}\;\, \\ 1\square4 \\ \times\,3 \\ \hline 582 \end{array} \qquad \begin{array}{r} {\scriptstyle 2\,1}\;\, \\ 1\boxed{9}4 \\ \times\,3 \\ \hline 582 \end{array}$$

- Diagram 2: Since $4 \times 3 = 12$, there is a 1 above the box.
- Diagram 3: Since $1 \times 3 = 3$ and $3 + 2 = 5$, there is a 2 above the hundreds digit.
- Diagram 4: $\square \times 3 + 1 = 28$. The missing digit equals 9 because $9 \times 3 + 1 = 27 + 1 = 28$.

We determined that the missing digit equals 9 by thinking through the multiplication process. When you finish, you can check the answer using the multiplication process. In this example, we get $4 \times 3 = 12$, $9 \times 3 + 1 = 27 + 1 = 28$, and $1 \times 3 + 2 = 3 + 2 = 5$, confirming that $194 \times 3 = 582$.

Problems. Determine the missing digits.

①

$$\begin{array}{r} 584 \\ \times\ 6 \\ \hline \square 504 \end{array}$$

②

$$\begin{array}{r} \square 95 \\ \times\ 8 \\ \hline 2360 \end{array}$$

③

$$\begin{array}{r} 3\square 7 \\ \times\ 4 \\ \hline 1388 \end{array}$$

④

$$\begin{array}{r} 942 \\ \times\ \square \\ \hline 6594 \end{array}$$

⑤

$$\begin{array}{r} \square 59 \\ \times\ 53 \\ \hline \square 77 \\ 12{,}950 \\ \hline 13{,}727 \end{array}$$

⑥

$$\begin{array}{r} 936 \\ \times\ \square 5 \\ \hline 4{,}680 \\ 7\square{,}880 \\ \hline 79{,}560 \end{array}$$

⑦

$$\begin{array}{r} 4\square 8 \\ \times\ 62 \\ \hline 816 \\ 24{,}\square 80 \\ \hline 25{,}296 \end{array}$$

⑧

$$\begin{array}{r} \square 74 \\ \times\ 49 \\ \hline 3{,}366 \\ 1\square{,}960 \\ \hline 18{,}326 \end{array}$$

1.7 Estimate by rounding one number

When multiplying by a one-digit number, one quick way to estimate the answer is to round the other number so that it has one digit followed by zeroes. For example, 385×7 can be rounded to 400×7. Multiply the nonzero digits together and add the same number of zeroes. For example, $400 \times 7 = 2800$. (This is 28 followed by two zeroes.) Compare 2800 to the actual value: $385 \times 7 = 2695$.

Examples. Estimate the answers by rounding.

(A) $62 \times 3 \approx 60 \times 3 = 180$ ($\approx$ means "approximately equal")

(B) $391 \times 6 \approx 400 \times 6 = 2400$

Problems. Estimate the answers by rounding.

① $18 \times 4 \approx$ ② $213 \times 8 \approx$

③ $31 \times 9 \approx$ ④ $687 \times 3 \approx$

⑤ $49 \times 6 \approx$ ⑥ $508 \times 7 \approx$

⑦ $67 \times 8 \approx$ ⑧ $819 \times 5 \approx$

⑨ $22 \times 3 \approx$ ⑩ $994 \times 4 \approx$

1.8 Estimate by rounding two numbers

When both numbers have multiple digits, one quick way to estimate the answer is to round both numbers so that they have one digit followed by zeroes. For example, 208×39 can be rounded to 200×40. Multiply the nonzero digits together and add the same number of zeroes. For example, $200 \times 40 = 8000$. (This is 8 followed by three zeroes.) Compare 8000 to the actual value: $208 \times 39 = 8112$.

Examples. Estimate the answers by rounding.

(A) $62 \times 3 \approx 60 \times 3 = 180$

(B) $391 \times 7 \approx 400 \times 7 = 2800$

Problems. Estimate the answers by rounding.

① $42 \times 31 \approx$

② $296 \times 41 \approx$

③ $79 \times 63 \approx$

④ $811 \times 18 \approx$

⑤ $39 \times 52 \approx$

⑥ $189 \times 49 \approx$

⑦ $63 \times 87 \approx$

⑧ $504 \times 78 \approx$

⑨ $98 \times 29 \approx$

⑩ $413 \times 88 \approx$

1.9 Multiplication in word problems

Multiplication is involved when the same number repeats. For example, if Sarah earns \$400 per week for 6 weeks, the value \$400 is repeated (it is paid 6 different times). To find Sarah's earnings for 6 weeks, multiply: $\$400 \times 6 = \$2400$.

Example. (A) A bookcase has 5 shelves. If each shelf holds 42 books, how many books does the bookcase hold?

Multiply 42 books by 5:

$$\begin{array}{r} {\scriptstyle 1} \\ 42 \\ \times\, 5 \\ \hline 210 \end{array}$$

The bookcase holds 210 books.

Problems. Solve each word problem.

① Sam bought 6 dozen eggs. How many eggs did Sam buy?

② How many balls are in 8 jars if each jar contains 73 balls?

③ Alex earned $39 every day for 24 days. How much money did Alex earn all together?

④ How many magnets are contained in 5 large boxes and 7 small boxes if each large box has 61 magnets and each small box has 26 magnets?

⑤ Pat has 48 blocks that weigh 14 pounds each and 36 blocks that weigh 25 pounds each. What is the total weight of all of the blocks?

1.10 Multiplying with parentheses

Do math inside parentheses first. For example, $5 \times (2 + 4)$ means to add 2 to 4 before multiplying by 5:

$$5 \times (2 + 4) = 5 \times 6 = 30$$

Examples. (A) $3 \times (1 + 5) = 3 \times 6 = 18$

(B) $(3 + 4) \times (9 - 4) = 7 \times 5 = 35$

Problems. Determine the answer to each problem.

① $8 \times (5 + 4) =$

② $(12 - 8) \times 6 =$

③ $3 \times (15 - 8) =$

④ $(2 + 3 + 4) \times 6 =$

⑤ $(5 + 2) \times (6 + 4) =$

⑥ $(3 + 3) \times (9 - 5) =$

⑦ $(14 - 6) \times (6 - 2) =$

⑧ $(13 - 8) \times (7 + 1) =$

1.11 Multiplying with brackets

Do math inside parentheses first. Do math inside brackets next. For example, $5 \times [(3 \times 2) - (20 \div 5)]$ means:

$$5 \times [(3 \times 2) - (20 \div 5)] = 5 \times [6 - 4] = 5 \times 2 = 10$$

Examples. (A) $2 \times [(4 \times 3) + 8] = 2 \times [12 + 8] = 2 \times 20 = 40$

(B) $[11 - (2 \times 4)] \times 6 = [11 - 8] \times 6 = 3 \times 6 = 18$

Problems. Determine the answer to each problem.

① $4 \times [2 + (12 \div 3)] =$

② $[(5 \times 3) - 8] \times 7 =$

③ $(14 - 9) \times [(32 \div 8) + 4] =$

④ $8 \times [(7 \times 3) - (6 \times 2)] =$

⑤ $[(9 \times 3) - 20] \times [30 - (4 \times 6)] =$

1.12 Multiplying without parentheses

The custom is to perform multiplication and division before addition and subtraction. (However, if you see parentheses, do math in parentheses first.) Work left to right when you multiply and divide. Then work left to right when you add and subtract. For example, $7 + 3 \times 2$ equals $7 + 6$ because we need to multiply before we add.

Examples. (A) $9 - 4 \times 2 = 9 - 8 = 1$ (multiply first)

(B) $12 \div 3 + 5 \times 2 = 4 + 10 = 14$ (divide and multiply first)

Problems. Determine the answer to each problem.

① $8 + 7 \times 6 =$

② $5 \times 4 - 3 =$

③ $15 - 6 \times 2 =$

④ $4 \times 7 - 2 \times 8 =$

⑤ $5 + 4 \times 2 + 7 =$

⑥ $180 \div 3 - 3 \times 4 \times 5 =$

1.13 Multiplication properties

According to the **commutative** property of multiplication, the order of the numbers doesn't matter. For example, 5×2 and 2×5 both equal 10. We can express the commutative property symbolically as:

$$a \times b = b \times a$$

The **associative** property of multiplication states that when three numbers are multiplied, the grouping doesn't matter. For example, $2 \times (3 \times 4) = 2 \times 12 = 24$ has the same answer as $(2 \times 3) \times 4 = 6 \times 4 = 24$. We can express the associative property symbolically as:

$$a \times (b \times c) = (a \times b) \times c$$

The **distributive** property involves multiplying one number by a sum of two numbers, like $5 \times (2 + 3)$. The distributive property states that $5 \times (2 + 3)$ is equal to $5 \times 2 + 5 \times 3$. If you multiply each of these out, you get $5 \times (2 + 3) = 5 \times 5 = 25$ and $5 \times 2 + 5 \times 3 = 10 + 15 = 25$, which both equal 25. We can express the distributive property symbolically as:

$$a \times (b + c) = a \times b + a \times c$$

According to the **identity** property of multiplication, any number times one is equal to itself. For example, $5 \times 1 = 5$. We can express the identity property symbolically as:

$$a \times 1 = 1 \times a = a$$

Examples. Apply the distributive property.

(A) $5 \times 16 = 5 \times (10 + 6) = 5 \times 10 + 5 \times 6 = 50 + 30 = 80$

(B) $48 \times 3 = (50 - 2) \times 3 = 50 \times 3 - 2 \times 3 = 150 - 6 = 144$

Problems. Apply the distributive property.

① $7 \times 63 =$

② $4 \times 71 =$

③ $9 \times 82 =$

④ $2 \times 39 =$

⑤ $6 \times 88 =$

⑥ $46 \times 8 =$

⑦ $57 \times 5 =$

⑧ $99 \times 9 =$

Multiple Choice Questions

① Which expression equals 42?

(A) 7×6 (B) 7×7 (C) 8×4 (D) 8×5 (E) 8×6

② What is 9×60?

(A) 54 (B) 480 (C) 540 (D) 4800 (E) 5400

③ What is 40×700?

(A) 2800 (B) 24,000 (C) 28,000 (D) 240,000 (E) 280,000

④ What is 87×6?

(A) 90 (B) 422 (C) 432 (D) 482 (E) 522

⑤ What is 493×7?

(A) 2831 (B) 2851 (C) 3431 (D) 3451 (E) 3493

⑥ What is 74×48?

(A) 888 (B) 3452 (C) 3552 (D) 3626 (E) 5032

⑦ What is 736×25?

(A) 1472 (B) 1840 (C) 5152 (D) 14,720 (E) 18,400

⑧ What is the missing digit in $9\square2 \times 8 = 7536$?

(A) 4 (B) 5 (C) 6 (D) 7 (E) 8

⑨ What are the missing digits in $2\square5 \times 37 = 10,\square45$?

(A) 7 and 5 (B) 7 and 6 (C) 8 and 5 (D) 8 and 6 (E) 9 and 6

⑩ Estimate 71×4.

(A) 240 (B) 280 (C) 320 (D) 360 (E) 400

⑪ Estimate 61×39.

(A) 180 (B) 240 (C) 1200 (D) 1800 (E) 2400

⑫ How many beads are in 6 jars if each jar has 73 beads?

(A) 428 (B) 438 (C) 458 (D) 488 (E) 498

⑬ A red jar has 32 beads. A blue jar has 24 beads. How many beads are in 5 red jars and 4 blue jars?

(A) 236 (B) 242 (C) 246 (D) 248 (E) 256

⑭ What is $20 - 6 \times 3$?

(A) 2 (B) 12 (C) 38 (D) 42 (E) 60

⑮ What is $8 \times [(9 \times 6) - (7 \times 7)]$?

(A) 24 (B) 32 (C) 40 (D) 48 (E) 56

⑯ What is $7 + 3 \times 2$?

(A) 12 (B) 13 (C) 19 (D) 20 (E) 35

⑰ The commutative property is illustrated by:

(A) $7 \times 1 = 7$ (B) $9 \times 5 = 5 \times 9$ (C) $4 \times (2 + 3) = 4 \times 2 + 4 \times 3$

⑱ The associative property is illustrated by:

(A) $2 \times (3 \times 4) = (2 \times 3) \times 4$ (B) $6 \times 9 = 9 \times 6$ (C) $4 \times 0 = 0$

⑲ Which expression is equivalent to $(60 \times 4) + (7 \times 4)$?

(A) 64×7 (B) 67×4 (C) 240×7 (D) 280×4 (E) 420×4

-2-

DIVISION

2.1 Quick review of division facts

Fluency with arithmetic facts is essential in math.

Problems. Practice these division facts.

① $24 \div 6 =$	② $18 \div 3 =$	③ $64 \div 8 =$
④ $36 \div 4 =$	⑤ $63 \div 9 =$	⑥ $25 \div 5 =$
⑦ $28 \div 7 =$	⑧ $30 \div 6 =$	⑨ $12 \div 3 =$
⑩ $18 \div 2 =$	⑪ $35 \div 7 =$	⑫ $32 \div 4 =$
⑬ $36 \div 6 =$	⑭ $48 \div 8 =$	⑮ $24 \div 3 =$
⑯ $81 \div 9 =$	⑰ $16 \div 4 =$	⑱ $42 \div 7 =$
⑲ $21 \div 7 =$	⑳ $30 \div 5 =$	㉑ $20 \div 5 =$
㉒ $40 \div 5 =$	㉓ $18 \div 6 =$	㉔ $14 \div 2 =$
㉕ $56 \div 8 =$	㉖ $25 \div 5 =$	㉗ $54 \div 6 =$
㉘ $27 \div 3 =$	㉙ $49 \div 7 =$	㉚ $72 \div 9 =$

2.2 Dividing by one digit

Follow these steps to divide by a one-digit divisor:

$3\overline{)192}$	3 doesn't divide into 1. Write nothing above the 1.
$\begin{array}{r} 6 \\ 3\overline{)192} \\ 18 \end{array}$	Find the biggest multiple of 3 not greater than 19. $3 \times 6 = 18$ doesn't exceed 19. Write 6 above the 9. Write 18 below the 19.
$\begin{array}{r} 64 \\ 3\overline{)192} \\ \underline{18} \\ 12 \end{array}$	$19 - 18 = 1$. Write 1 below the 9. Bring down the 2. Which multiple of 3 equals 12? $3 \times 4 = 12$. Write 4 above the 2.

The final answer is 64.

Here is another way of looking at it:

- $192 \div 3$ lies between 60 and 70 since $3 \times 60 = 180$ and $3 \times 70 = 210$.
- $192 - 180 = 12$.
- $12 \div 3 = 4$.
- $192 \div 3 = 64$ because $3 \times 60 = 180$ and $3 \times 4 = 12$.

Check the answer by multiplying: $64 \times 3 = 192$.

Examples. Divide the numbers.

(A) $3\overline{)87}$ = 29

$$\begin{array}{r} 29 \\ 3\overline{)87} \\ \underline{6} \\ 27 \end{array}$$

$3 \times 2 = 6$ doesn't exceed 8. Write 2 above the 8. Write 6 below the 8. Subtract 6 from 8. Bring down the 7.
$3 \times 9 = 27$ Write 9 above the 7.

(B)

$$\begin{array}{r} 87 \\ 5\overline{)435} \\ \underline{40} \\ 35 \end{array}$$

5 doesn't divide into 4. Write nothing above the 4.
$5 \times 8 = 40$ doesn't exceed 43. Write 8 above the 3. Write 40 below the 43. Subtract 40 from 43. Bring down the 5.
$5 \times 7 = 35$ Write 7 above the 5.

(C)

$$\begin{array}{r} 241 \\ 4\overline{)964} \\ \underline{8} \\ 16 \\ \underline{16} \\ 04 \end{array}$$

$4 \times 2 = 8$ doesn't exceed 9. Write 2 above the 9. Write 8 below the 9. Subtract 8 from 9. Bring down the 6.
$4 \times 4 = 16$ doesn't exceed 16. Write 4 above the 6. Write 16 below the 16. Subtract 16 from 16. Bring down the 4.
$4 \times 1 = 4$ Write 1 above the 4.

Problems. Divide the numbers.

① $6\overline{)84}$

② $8\overline{)576}$

③
$2\overline{)922}$

④
$5\overline{)315}$

⑤
$7\overline{)343}$

⑥
$3\overline{)4284}$

⑦
$4\overline{)952}$

⑧
$9\overline{)5148}$

2.3 Visualizing division

We can draw pictures to visualize a division problem.

- A tiny circle represents a single unit.
- A line represents a ten. (It is 10 units long.)
- A square represents a hundred. (It is 10×10.)

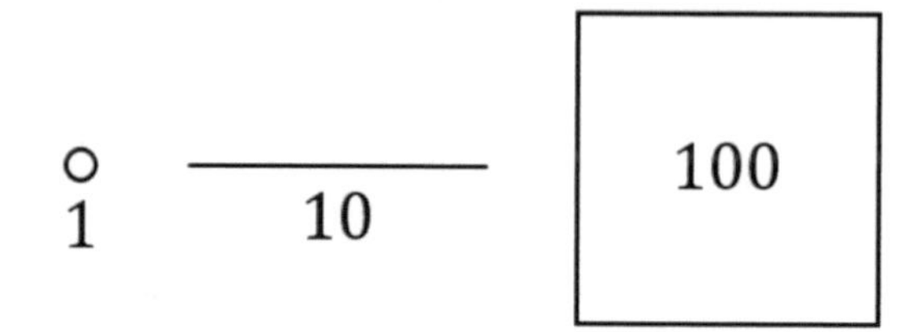

These are called **base-ten blocks**. We can use base-ten blocks to draw a division problem following these steps:

- First draw squares, lines, and tiny circles to add up to the dividend. For example, for $378 \div 14$, the dividend (378) would be drawn as 3 squares, 7 lines, and 8 circles.
- Rearrange the squares, lines, and tiny circles to make vertical groups equal to the divisor. For $378 \div 14$, the divisor (14) would be drawn as 1 line and 4 circles.
- It is usually necessary to regroup. A line can be redrawn as 10 tiny circles. A square can be redrawn as 10 lines.
- Count the number of divisors going across. This equals the answer to the division problem.

An example of how to draw 378 ÷ 14 is illustrated below.

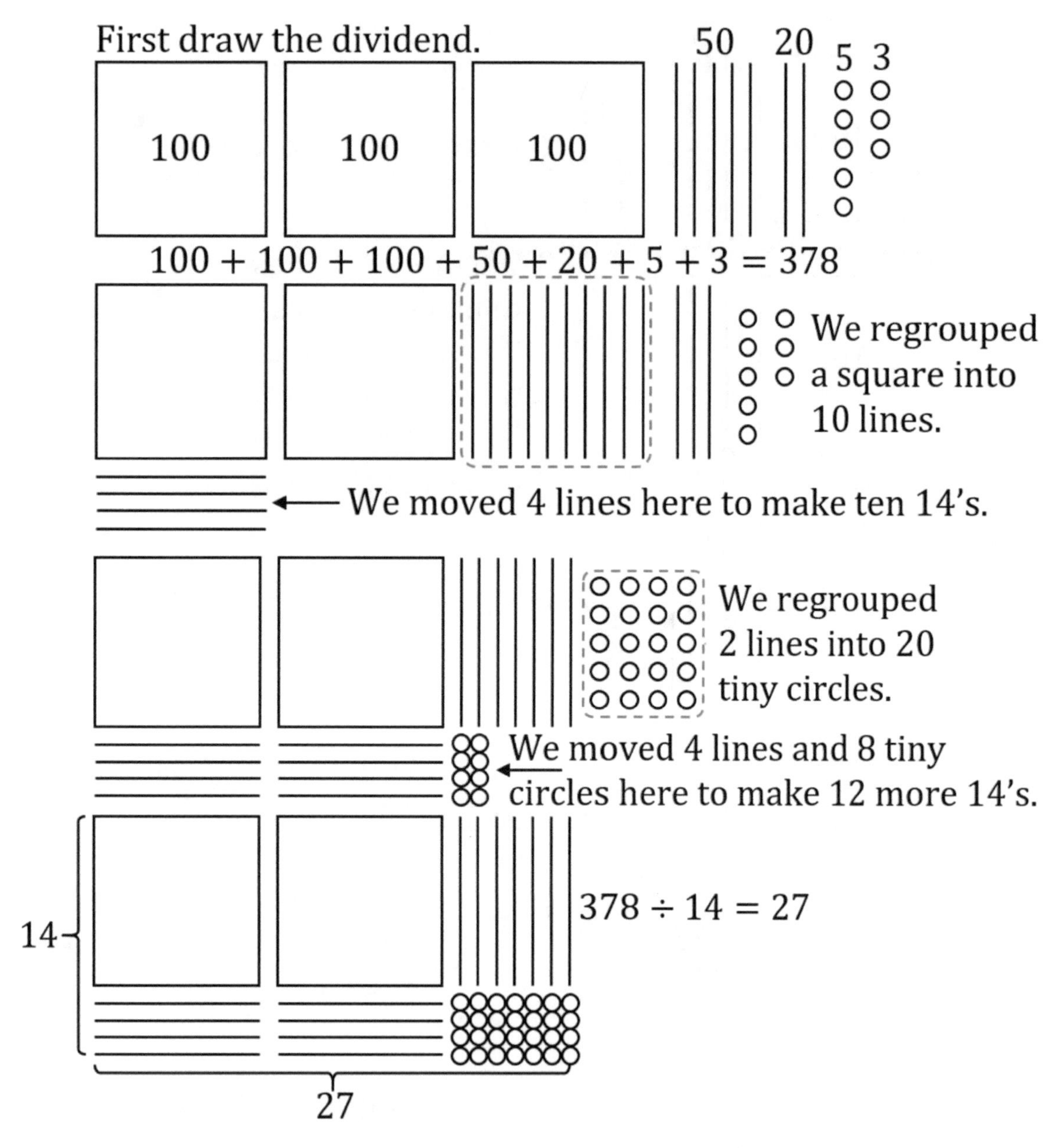

In the second step, we regrouped 1 square into 10 lines, and in the third step, we regrouped 2 lines into 2 sets of 10 tiny circles. The answer has 27 groups of 14.

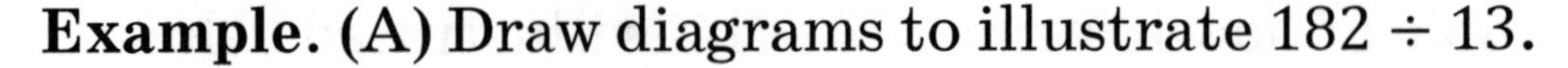

Example. (A) Draw diagrams to illustrate 182 ÷ 13.

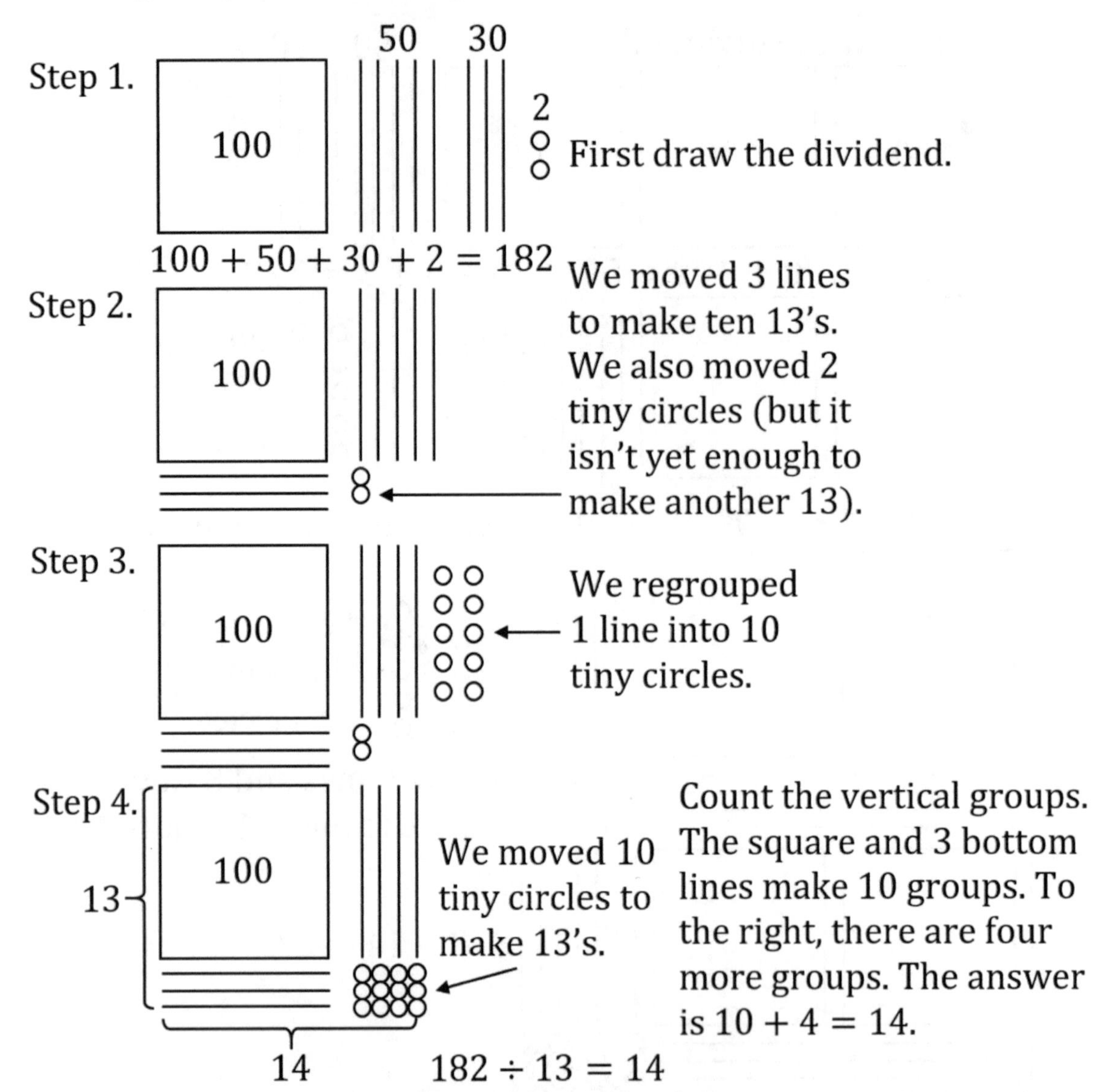

Step 1.
50
30
100
2
First draw the dividend.
100 + 50 + 30 + 2 = 182
Step 2.
100
8
We moved 3 lines to make ten 13's. We also moved 2 tiny circles (but it isn't yet enough to make another 13).
Step 3.
100
8
We regrouped 1 line into 10 tiny circles.
Step 4.
13
100
We moved 10 tiny circles to make 13's.
Count the vertical groups. The square and 3 bottom lines make 10 groups. To the right, there are four more groups. The answer is 10 + 4 = 14.
14
182 ÷ 13 = 14

The final picture is a rectangle. The height of the rectangle is 13 (that was the divisor). The width of the rectangle is 14 (that is the answer, called the quotient). You can check the answer by multiplying: 14 × 13 = 182.

Problems. Draw diagrams to solve each problem visually.

① $192 \div 12 =$

(A) Draw the dividend.

(B) Redraw the picture with regrouping.

(C) Draw the final picture showing the answer.

(D) According to the final picture, what is $192 \div 12$?

② $253 \div 23 =$

(A) Draw the dividend.

(B) Redraw the picture with regrouping.

(C) Draw the final picture showing the answer.

(D) According to the final picture, what is $253 \div 23$?

2.4 Dividing by two digits

Dividing by a two-digit divisor is very similar to dividing by a one-digit divisor (Sec. 2.2):

$$\begin{array}{r} 5 \\ 15\overline{)795} \\ 75 \end{array}$$

Find the biggest multiple of 15 not greater than 79.
$15 \times 5 = 75$ doesn't exceed 79. Write 5 above the 9.
Write 75 below the 79.

$$\begin{array}{r} 53 \\ 15\overline{)795} \\ \underline{75} \\ 45 \end{array}$$

$79 - 75 = 4$. Write 4 below the 75.
Bring down the 5.
Which multiple of 15 equals 45?
$15 \times 3 = 45$.
Write 3 above the 5.

The final answer is 53.

Here is another way of looking at it:

- $795 \div 15$ lies between 750 and 900 since $15 \times 50 = 750$ and $15 \times 60 = 900$.
- $795 - 750 = 45$.
- $45 \div 15 = 3$.
- $795 \div 15 = 53$ because $15 \times 50 = 750$ and $15 \times 3 = 45$.

Check the answer by multiplying: $53 \times 15 = 795$.

Examples. Divide the numbers.

(A)
$$\begin{array}{r} 8 \\ 12\overline{)96} \\ \underline{96} \\ 0 \end{array}$$

$12 \times 8 = 96$ doesn't exceed 96. Write 8 above the 6. Write 96 below the 96. Subtract 96 from 96. Since this equals zero (and there aren't any digits remaining), we are finished.

(B)
$$\begin{array}{r} 35 \\ 25\overline{)875} \\ \underline{75} \\ 125 \end{array}$$

$25 \times 3 = 75$ doesn't exceed 87. Write 3 above the 7. Write 75 below the 87. Subtract 75 from 87. Bring down the 5.
$25 \times 5 = 125$ Write 5 above the 5.

(C)
$$\begin{array}{r} 6 \\ 72\overline{)432} \\ \underline{432} \\ 0 \end{array}$$

72 doesn't divide into 43. Write nothing above the 3. $72 \times 6 = 432$ doesn't exceed 432. Write 6 above the 2. Write 432 below the 432. Subtract 432 from 432. Since this equals zero (and there aren't any digits remaining), we are finished.

Problems. Divide the numbers.

① $14\overline{)98}$

② $32\overline{)864}$

③

$51\overline{)714}$

④

$23\overline{)966}$

⑤

$87\overline{)957}$

⑥

$43\overline{)731}$

⑦

$18\overline{)846}$

⑧

$79\overline{)632}$

2.5 Four digits divided by two digits

The dividends in this section have four digits. Recall that the **dividend** divided by the **divisor** equals the **quotient**. For example, in $12 \div 3 = 4$, the dividend is 12, the divisor is 3, and the quotient is 4.

Examples. Divide the numbers.

(A)

$$\begin{array}{r} 58 \\ 25\overline{)1450} \\ \underline{125} \\ 200 \end{array}$$

25 doesn't divide into 14. Write nothing above the 4.
25 × 5 = 125 doesn't exceed 145.
Write 5 above the 5. Write 125 below the 145.
Subtract 125 from 145. Bring down the 0.
25 × 8 = 200 Write 8 above the 0.

(B)

$$\begin{array}{r} 42 \\ 83\overline{)3486} \\ \underline{332} \\ 166 \end{array}$$

83 doesn't divide into 34. Write nothing above the 4.
83 × 4 = 332 doesn't exceed 348.
Write 4 above the 8. Write 332 below the 348.
Subtract 332 from 348. Bring down the 6.
83 × 2 = 166 Write 2 above the 6.

Problems. Divide the numbers.

① $36\overline{)2412}$

② $50\overline{)4450}$

③

$27\overline{)1458}$

④

$96\overline{)5952}$

⑤

$18\overline{)1656}$

⑥

$65\overline{)2210}$

⑦

$41\overline{)3157}$

⑧

$83\overline{)3984}$

2.6 Missing digit challenge

The division problems in this section are solved, but one of the digits is missing from the problem or solution. You can reason out the missing digit by thinking through the steps involved in the division process.

Examples. Determine the missing digit.

(A)
$$\begin{array}{r} \boxed{6}7 \\ 4\overline{)268} \\ \underline{24} \\ 28 \end{array}$$

(B)
$$\begin{array}{r} 29 \\ 17\overline{)4\boxed{9}3} \\ \underline{34} \\ 153 \end{array}$$

- **(A)** $4 \times ? = 24$ implies that the missing digit is 6 since $4 \times 6 = 24$.
- **(B)** $4? - 34 = 15$ implies that the missing digit is 9 since $49 - 34 = 15$.

Problems. Determine the missing digits.

①
$$\begin{array}{r} \boxed{}9 \\ 8\overline{)632} \\ \underline{56} \\ 72 \end{array}$$

②
$$\begin{array}{r} 17 \\ 43\overline{)7\boxed{}1} \\ \underline{43} \\ 301 \end{array}$$

③
$$\begin{array}{r} 573 \\ 6\overline{)3\boxed{}38} \\ \underline{30} \\ 43 \\ \underline{42} \\ 18 \end{array}$$

2.7 Partial quotients

The method of **partial quotients** is an alternative division strategy. This method is less efficient, but allows students to work with simpler multiples, like 10 times the divisor. Subtract easy multiples of the divisor until reaching zero. Write the number multiplied in a column on the right. Add up the numbers on the right to get the answer.

32		
43)1376		
−430	10	We multiplied 43 by 10.
946		Subtract: $1376 - 430 = 946$
−430	10	We multiplied 43 by 10 again.
516		Subtract: $946 - 430 = 516$
−430	10	We multiplied 43 by 10 again.
86		Subtract: $516 - 430 = 86$
−86	+2	We multiplied 43 by 2.
0	32	Add the numbers in the right column: $10 + 10 + 10 + 2 = 32$ is the answer.

In the example of partial quotients above, we multiplied 43 by 10 three times and multiplied 43 by 2 once. Add the three 10's and 2 to find the answer: $10 + 10 + 10 + 2 = 32$. Note that the method from Sec.'s 2.4-2.5 would solve the problem with less work, as shown on the following page.

$$\begin{array}{r} 32 \\ 43\overline{)1376} \\ \underline{129} \\ 86 \end{array}$$

As shown above, it simpler to multiply 43 by 30 and then multiply 43 by 2 than to multiply by 10, 10, 10, and 2. Both methods give the same answer (32).

Example. Divide the numbers using partial quotients.

(A)

$$\begin{array}{rrl} 27 & & \\ 56\overline{)1512} & & \\ -\underline{560} & 10 & \text{We multiplied 56 by 10.} \\ 952 & & \text{Subtract: } 1512 - 560 = 952 \\ -\underline{560} & 10 & \text{We multiplied 56 by 10 again.} \\ 392 & & \text{Subtract: } 952 - 560 = 392 \\ -\underline{280} & 5 & \text{We multiplied 56 by 5. (10 is too big.)} \\ 112 & & \text{Subtract: } 392 - 280 = 112 \\ -\underline{112} & \underline{+2} & \text{We multiplied 56 by 2.} \\ 0 & 32 & \text{Add the numbers in the right column:} \\ & & 10 + 10 + 5 + 2 = 27 \text{ is the answer.} \end{array}$$

Problems. Divide the numbers using your preferred method.

① $19\overline{)437}$

② $74\overline{)2294}$

③

$67\overline{)2211}$

④

$43\overline{)1118}$

⑤

$58\overline{)928}$

⑥

$81\overline{)1377}$

⑦

$24\overline{)648}$

⑧

$96\overline{)3936}$

2.8 Remainders

When the divisor doesn't evenly divide into the dividend, there is a **remainder**. As a simple example, $23 \div 4 = 5\ R3$. This means 5 with a remainder of 3. There is a remainder because 4 doesn't evenly divide into 23. You can divide 4 into 20 since $20 \div 5 = 4$, and the remainder is $23 - 20 = 3$.

Examples. Divide the numbers and find the remainder.

(A)

$$\begin{array}{r} 34\ R11 \\ 25\overline{)861} \\ \underline{75} \\ 111 \\ \underline{100} \\ 11 \end{array}$$

$25 \times 3 = 75$ is as close as we can get to 86.

$25 \times 4 = 100$ is as close as we can get to 111.

The remainder is 11. Write $R11$ at the top.

(B)

$$\begin{array}{r} 82\ R6 \\ 14\overline{)1154} \\ \underline{112} \\ 34 \\ \underline{28} \\ 6 \end{array}$$

$14 \times 8 = 112$ is as close as we can get to 115.

$14 \times 2 = 28$ is as close as we can get to 34.

The remainder is 6. Write $R6$ at the top.

The answer to (A) is $34\ R11$, meaning 34 with a remainder of 11. Check the answer: $34 \times 25 = 850$ and $861 - 850 = 11$. The answer to (B) is $82\ R6$, meaning 82 with a remainder of 6. Check the answer: $82 \times 14 = 1148$ and $1154 - 1148 = 6$.

Problems. Divide the numbers and find the remainder.

①

$37\overline{)892}$

②

$63\overline{)1147}$

③

$45\overline{)3022}$

④

$92\overline{)3957}$

⑤

$78\overline{)2761}$

⑥

$54\overline{)4657}$

2.9 Estimate division

A quick way to estimate the answer to a division problem is to add zeroes to a suitable division fact. As an example, consider $1601 \div 29$. We can estimate the answer by adding zeroes to $15 \div 3$ to get $1500 \div 30$.

- $15 \div 3 = 5$
- $150 \div 3 = 50$
- $1500 \div 3 = 500$
- $1500 \div 30 = 50$

Adding a zero to the dividend makes the answer 10 times larger, while adding a zero to the divisor makes the answer 10 times smaller. We may use $1500 \div 30 = 50$ to estimate 50 as the answer to $1601 \div 29$. For comparison, the exact answer is $1601 \div 29 = 55\ R6$.

When estimating, it is useful to multiply in order to check the estimate. Our estimate was $1500 \div 30 = 50$. To check this, multiply $50 \times 30 = 1500$ (which is 15 followed by two zeroes).

Examples. Estimate the answers by rounding.

(A) $109 \div 6 \approx 120 \div 6 = 20$ Check: $20 \times 6 = 120$

(B) $1303 \div 31 \approx 1200 \div 30 = 40$ Check: $40 \times 30 = 1200$

(C) $7895 \div 39 \approx 8000 \div 40 = 200$ Check: $200 \times 40 = 8000$

Problems. Estimate the answers by rounding.

① $418 \div 7 \approx$

② $573 \div 8 \approx$

③ $8373 \div 21 \approx$

④ $2912 \div 42 \approx$

⑤ $2345 \div 29 \approx$

⑥ $1536 \div 49 \approx$

⑦ $499 \div 72 \approx$

⑧ $4428 \div 88 \approx$

⑨ $5294 \div 59 \approx$

⑩ $3271 \div 81 \approx$

2.10 Division in word problems

Division is involved when a quantity is divided into equal portions. For example, to distribute 300 gumballs equally among 5 people, divide: $300 \div 5 = 60$.

Example. (A) A farmer wishes to plant 180 seeds in 12 rows. How many seeds should the farmer plant in each row?

Divide 180 seeds by 12:

$$\begin{array}{r} 15 \\ 12\overline{)180} \\ \underline{12} \\ 60 \end{array}$$

The farmer should plant 15 seeds in each row.

Problems. Solve each word problem.

① A group of 32 students has 1248 pencils. If the pencils are divided equally among the students, how many pencils will each student receive?

② Patty read twelve times as many pages as Sally. Together they read 312 pages. How many pages did each person read?

③ At a movie theater, one adult ticket costs twice as much as one children's ticket. A mother pays a total of $18 to buy one ticket for herself and another ticket for her daughter. How much does each ticket cost?

④ A jar can hold up to 87 balls. A total of 1000 balls are put into 12 jars. If all of the jars are full except for one jar, how many balls are in the jar that contains the fewest balls?

2.11 Division with parentheses

Do math inside parentheses first. For example, $(7 + 5) \div 3$ means to add 7 to 5 before dividing by 3:

$$(7 + 5) \div 3 = 12 \div 3 = 4$$

Examples. (A) $8 \div (3 + 1) = 8 \div 4 = 2$

(B) $(9 + 7) \div (7 - 3) = 16 \div 4 = 4$

Problems. Determine the answer to each problem.

① $(24 + 12) \div 9 =$

② $30 \div (11 - 6) =$

③ $(48 - 16) \div 4 =$

④ $35 \div (5 + 4 - 2) =$

⑤ $(9 + 8 + 7) \div 8 =$

⑥ $(36 + 36) \div (16 - 8) =$

⑦ $(30 - 9) \div (10 - 7) =$

⑧ $(24 + 16) \div (3 + 2) =$

2.12 Division with brackets

Do math inside parentheses first. Do math inside brackets next. For example, $[(8 \times 4) - (8 \div 4)] \div 5$ means:

$$[(8 \times 4) - (8 \div 4)] \div 5 = [32 - 2] \div 5 = 30 \div 5 = 6$$

Examples. (A) $28 \div [(12 \div 3) + 3] = 28 \div [4 + 3] = 28 \div 7 = 4$

(B) $[3 + (9 \div 3)] \div 3 = [3 + 3] \div 3 = 6 \div 3 = 2$

Problems. Determine the answer to each problem.

① $[9 + (54 \div 6)] \div 6 =$

② $[(7 \times 8) - 8] \div 8 =$

③ $36 \div [(24 \div 4) + 3] =$

④ $15 \div [(56 \div 7) - (25 \div 5)] =$

⑤ $[(7 \times 6) + 3] \div [13 - (16 \div 2)] =$

2.13 Division without parentheses

The custom is to perform multiplication and division before addition and subtraction. (However, if you see parentheses, do math in parentheses first.) Work left to right when you multiply and divide. Then work left to right when you add and subtract. For example, $6 + 9 \div 3$ equals $6 + 3$ because we need to divide before we add.

Examples. (A) $7 - 10 \div 5 = 7 - 2 = 5$ (divide first)

(B) $5 \times 4 - 21 \div 3 = 20 - 7 = 13$ (multiply and divide first)

Problems. Determine the answer to each problem.

① $4 + 6 \div 2 =$

② $24 \div 6 - 2 =$

③ $12 - 8 \div 4 =$

④ $3 \times 8 - 4 \div 2 =$

⑤ $9 + 6 \div 3 - 2 =$

⑥ $9 \times 8 \div 6 - 6 \div 2 =$

2.14 Divisibility tests

It is often useful to know whether a given number is evenly divisible by 2, 3, 4, 5, 6, or 9. The following rules can help you quickly find out if a number is evenly divisible by 2, 3, 4, 5, 6, or 9.

A number is evenly divisible by 2 if the number is even. For example, 36 is evenly divisible by 2 since 36 is even.

A number is evenly divisible by 3 if the digits add up to a multiple of 3. For example, 48 is evenly divisible by 3 since $4 + 8 = 12$ and since 12 is a multiple of 3.

A number is evenly divisible by 4 if the last two digits are a multiple of 4. For example, 216 is divisible by 4 because 16 is a multiple of 4.

A number is evenly divisible by 5 if the number ends with 0 or 5. For example, 55 is evenly divisible by 5 since 55 ends with 5 and 70 is evenly divisible by 5 since 70 ends with 0.

A number is evenly divisible by 6 if it passes the tests for both 2 and 3. For example, 72 is evenly divisible by 6 because 72 is even and because $7 + 2 = 9$ is a multiple of 3.

A number is evenly divisible by 9 if the digits add up to a multiple of 9. For example, 468 is evenly divisible by 9 since $4 + 6 + 8 = 18$.

Examples. (A) 375 is evenly divisible by 3 because $3 + 7 + 5 = 15$ is a multiple of 3 and 375 is evenly divisible by 5 because 375 ends with 5. (However, 375 isn't evenly divisible by 2 because 375 isn't even, 375 isn't evenly divisible by 4 because 75 isn't a multiple of 4, 375 isn't evenly divisible by 6 because 375 didn't pass the test for 2, and 375 isn't evenly divisible by 9 because $3 + 7 + 5 = 15$ isn't a multiple of 9.)

(B) 108 is evenly divisible by 2 because 108 is even, 108 is evenly divisible by 3 because $1 + 0 + 8 = 9$ is a multiple of 3, 108 is evenly divisible by 4 because 08 is a multiple of 4, 108 is evenly divisible by 6 because 108 passes the rules for 2 and 3, and 108 is evenly divisible by 9 because $1 + 0 + 8 = 9$ is a multiple of 9. (However, 108 isn't evenly divisible by 5 because it doesn't end with 0 or 5.)

Problems. For each number below, is the number evenly divisible by 2, 3, 4, 5, 6, or 9? List all that apply.

① 70

② 78

③ 220

④ 195

⑤ 66

⑥ 72

⑦ 180

⑧ 495

⑨ 700

⑩ 218,790

2.15 Greatest common factor

To find the **greatest common factor** (GCF) of two numbers, find the largest whole number that evenly divides into each given number. For example, 9 is the GCF of 27 and 36 since 9 is the largest whole number that evenly divides into both 27 and 36.

Examples. (A) What is the GCF of 18 and 24?

$18 = 6 \times 3$ and $24 = 6 \times 4$. The GCF is 6.

(B) What is the GCF of 25 and 35?

$25 = 5 \times 5$ and $35 = 5 \times 7$. The GCF is 5.

Problems. Determine the GCF for each pair of numbers.

① 8, 12 ② 9, 12

③ 24, 54 ④ 15, 20

⑤ 36, 63 ⑥ 35, 49

2.16 Least common multiple

To find the **least common multiple** (LCM) of two numbers, find the smallest whole number that is evenly divisible by each number. For example, 12 is the LCM of 4 and 6.

Examples. (A) What is the LCM of 2 and 3? List their multiples. 2, 4, [6], 8 ... and 3, [6], 9, 12 ... The LCM is 6.

(B) What is the LCM of 6 and 8? List their multiples.

6, 12, 18, [24], 30 ... and 8, 16, [24], 32 ... The LCM is 24.

(C) What is the LCM of 9 and 12? List their multiples.

9, 18, 27, [36], 45 ... and 12, 24, [36], 48 ... The LCM is 36.

Problems. Determine the LCM for each pair of numbers.

① 3, 4

② 8, 10

③ 6, 9

④ 10, 25

⑤ 5, 7

⑥ 12, 16

Multiple Choice Questions

① Which expression equals 7?

(A) $36 \div 6$ (B) $48 \div 6$ (C) $49 \div 9$ (D) $56 \div 8$ (E) $63 \div 7$

② What is $320 \div 8$?

(A) 4 (B) 35 (C) 40 (D) 350 (E) 400

③ What is $4200 \div 60$?

(A) 8 (B) 70 (C) 80 (D) 700 (E) 800

④ What is $702 \div 9$?

(A) 67 (B) 76 (C) 78 (D) 79 (E) 87

⑤ What is $476 \div 14$?

(A) 26 (B) 34 (C) 36 (D) 44 (E) 46

⑥ What is $1537 \div 53$?

(A) 29 (B) 31 (C) 33 (D) 35 (E) 37

⑦ What is the missing digit in $2\square73 \div 47 = 59$?

(A) 4 (B) 5 (C) 6 (D) 7 (E) 8

⑧ Estimate $2264 \div 38$.

(A) 40 (B) 50 (C) 60 (D) 400 (E) 600

⑨ What is the remainder for $2325 \div 86$?

(A) 1 (B) 2 (C) 3 (D) 4 (E) 5

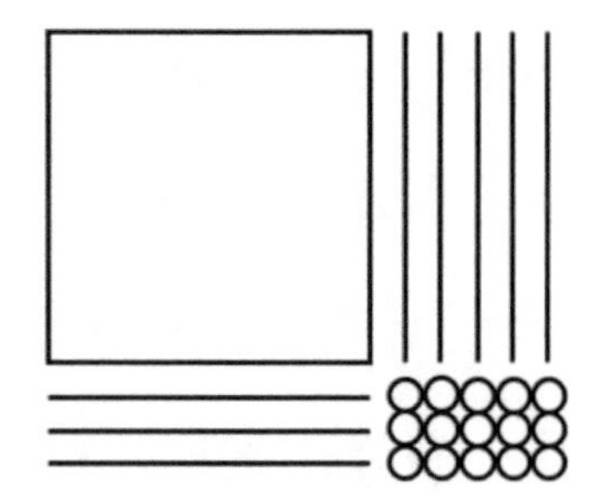

⑩ Which division problem is pictured above?

(A) $180 \div 3$ (B) $180 \div 11$ (C) $185 \div 13$

(D) $195 \div 11$ (E) $195 \div 13$

⑪ What is the answer to the problem in Exercise 10?

(A) 15 (B) 23 (C) 25 (D) 43 (E) 60

⑫ If 414 erasers are split equally among 18 students, how many erasers will each student receive?

(A) 18 (B) 19 (C) 21 (D) 22 (E) 23

⑬ Joe is 11 times as old as Amy. The sum of their ages is 84. How old is Joe?

(A) 44 (B) 55 (C) 66 (D) 77 (E) 88

⑭ What is $(72 \div 9) \div 4$?

(A) 1 (B) 2 (C) 2.5 (D) 4 (E) 32

⑮ What is $48 \div [(4 \times 3) - (18 \div 3)]$?

(A) 1 (B) 2 (C) 4 (D) 6 (E) 8

⑯ What is $6 + 12 \div 3$?

(A) 3 (B) 4 (C) 6 (D) 9 (E) 10

⑰ Which number is evenly divisible by 2, 3, and 5?

(A) 45 (B) 50 (C) 90 (D) 144 (E) 200

⑱ Which number is evenly divisible by 9?

(A) 4218 (B) 5435 (C) 5888 (D) 6213 (E) 7614

⑲ What is the greatest common factor of 32 and 56?

(A) 4 (B) 6 (C) 8 (D) 14 (E) 16

⑳ What is the least common multiple of 9 and 15?

(A) 24 (B) 36 (C) 45 (D) 72 (E) 135

-3-

DECIMAL PLACE VALUES

3.1 Place values

The **place value** of a digit indicates its position in a number. For example, for the number below, the following list gives the place value of each digit:

42,738.1569

- The 4 is in the ten thousands place.
- The 2 is in the thousands place.
- The 7 is in the hundreds place.
- The 3 is in the tens place.
- The 8 is in the units place.
- The 1 is in the **tenths** place.
- The 5 is in the **hundredths** place.
- The 6 is in the **thousandths** place.
- The 9 is in the **ten thousandths** place.

Read carefully because the "th" in "tenths," "hundredths," and "thousandths" is very significant. For example, in the number 378.16, the 6 is in the hundredths place whereas the 3 is in the hundreds place. Similarly, the 1 is in the tenths place whereas the 7 is in the tens place. Whether or not the word ends with "th" makes a big difference.

Examples. (A) What is the place value of each digit in 1.42?

The 1 is in the units place, the 4 is in the tenths place, and the 2 is in the hundredths place.

(B) What is the place value of the 7 in 0.0279?

The 7 is in the thousandths place.

Problems. What is the place value of the indicated digit?

① the 2 in 3.2

② the 4 in 46

③ the 5 in 0.25

④ the 8 in 813

⑤ the 7 in 0.007

⑥ the 1 in 1.9

⑦ the 6 in 2.486

⑧ the 3 in 53,178

⑨ the 9 in 5.97

⑩ the 0 in 1.9603

3.2 Expanded form of decimals

Consider the number 423, which is four hundred twenty-three. Observe that this number equals $400 + 20 + 3$. This is called the **expanded form** of the number. The same idea applies to decimals, where $\frac{1}{10}$ is a tenth, $\frac{1}{100}$ is a hundredth, $\frac{1}{1000}$ is a thousandth, etc. For example, in expanded form, the number 28.74 is:

$$28.74 = 20 + 8 + \frac{7}{10} + \frac{4}{100}$$

Examples. Write each number in expanded form.

(A) $217.3 = 200 + 10 + 7 + \frac{3}{10}$

(B) $0.49 = 0 + \frac{4}{10} + \frac{9}{100}$ or simply $\frac{4}{10} + \frac{9}{100}$

(C) $8.236 = 8 + \frac{2}{10} + \frac{3}{100} + \frac{6}{1000}$

Problems. Write each number in expanded form.

① 5.3

② 37.52

③ 417.8

④ 0.095

⑤ 2.837

⑥ 3812.1

⑦ 0.57

⑧ 29.463

⑨ 6.25

⑩ 2.5179

3.3 Writing decimals

When a decimal number is written in words:

- We use "and" for the decimal point.
- We write the digits following the decimal point like we would normally write the number (as if they weren't decimal digits).
- We end with the word for the last decimal place (and include an "s" at the end of it).

For example, the number 4.317 is "four and three hundred seventeen thousandths." We wrote "four" for 4, "and" for the decimal point, "three hundred seventeen" for 317, and the word "thousandths" for the place value of the 7 (which is the final digit).

Examples. Write each number in words.

(A) 15.3 is fifteen and three tenths

(B) 6.48 is six and forty-eight hundredths

(C) 7.196 is seven and one hundred ninety-six thousandths

Problems. Write each number in words.

① 7.2

② 8.14

③ 94.7

④ 3.758

⑤ 15.23

⑥ 8.059

⑦ 423.987

3.4 Place value charts

A place value chart has one column for each digit, with a decimal point between the units column and tenths column. For example, the chart below shows the number 27.985.

tens	units	.	tenths	hundredths	thousandths
2	7	.	9	8	5

Example. (A) Write 4.39 in the place value chart.

tens	units	.	tenths	hundredths	thousandths
	4	.	3	9	

Problems. Write each number in the place value chart.

① 5.314

tens	units	.	tenths	hundredths	thousandths
		.			

② 74.62

tens	units	.	tenths	hundredths	thousandths
		.			

③ 81.359

tens	units	.	tenths	hundredths	thousandths
		.			

3.5 Visualizing decimals

We can draw pictures to visualize a decimal value.

- A square represents a single unit. It is 1 by 1.
- The square can be divided into 10 strips to represent tenths. Each strip represents one tenth.
- The square can be divided into a 10×10 grid (with 100 tiny squares) to represent hundredths. Each tiny square represents one hundredth. A stack of 10 tiny squares represents one tenth.
- To represent a thousandth, one tiny square from the 10×10 grid can be magnified and divided into 10 strips. Each strip of this magnified tiny square represents a thousandth. See the right two figures below.

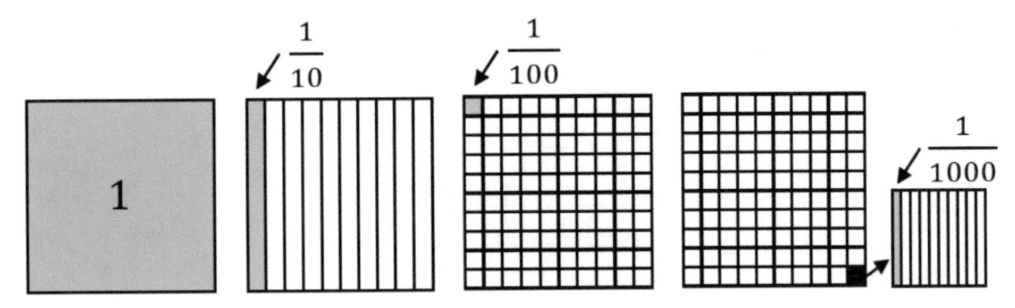

A decimal value can be represented by shading strips or tiny squares corresponding to the place values of the digits.

Examples. (A) Draw a picture to represent 0.7.

Shade 7 out of 10 strips gray.

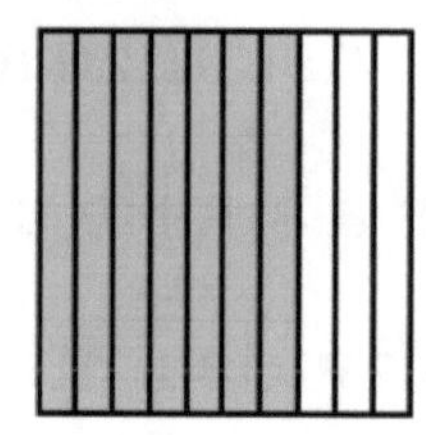

(B) Draw a picture to represent 0.38.

Shade 3 out of 10 strips gray plus 8 tiny squares.

Together, these represent 38 out of 100 tiny squares.

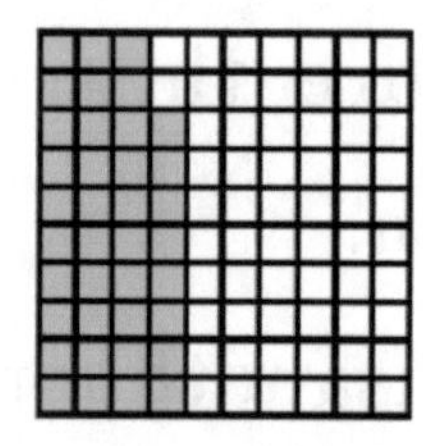

(C) Draw a picture to represent 0.826.

Shade 8 out of 10 strips gray plus 2 tiny squares plus 6 strips of a magnified tiny square. The 6 strips of the magnified tiny square represent 6 thousandths.

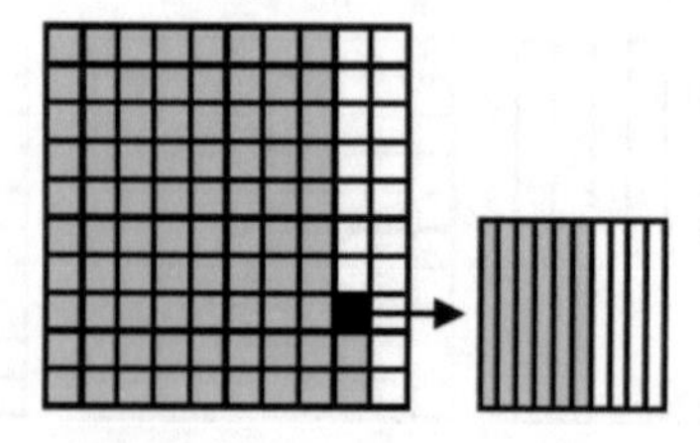

Problems. Shade the square gray according to the decimal. If needed, add a magnified tiny square.

① 0.4

② 0.68

③ 0.374

3.6 Decimals on the number line

A number line helps to visualize the order of numbers. For example, the number line below shows decimal values up to 1 in increments of 0.1.

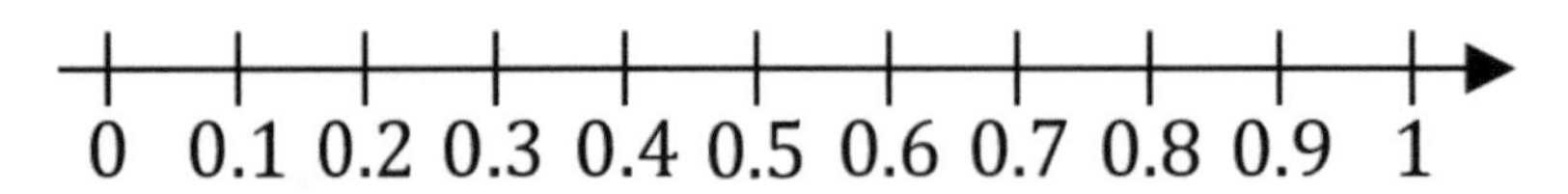

Example. (A) Draw and label 0.27 on the number line.

0.27 is closer to 0.3 than 0.2 because it is greater than 0.25.

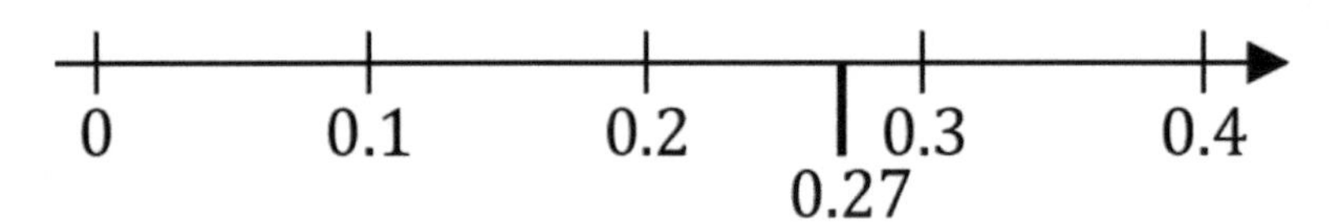

Problems. Draw and label each value on the number line.

① 0.02 ② 0.15 ③ 0.21 ④ 0.38

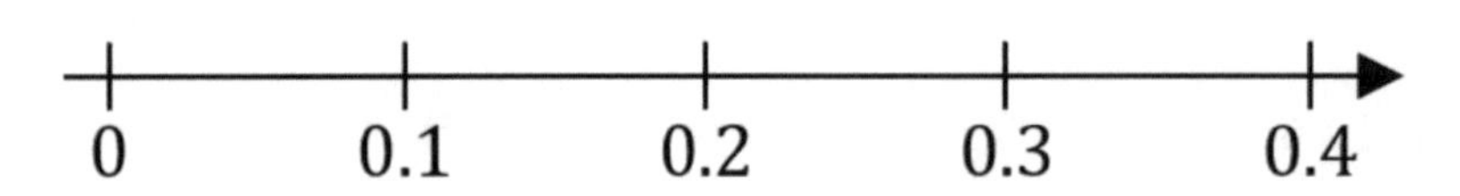

⑤ 0.22 ⑥ 0.47 ⑦ 0.65 ⑧ 0.83

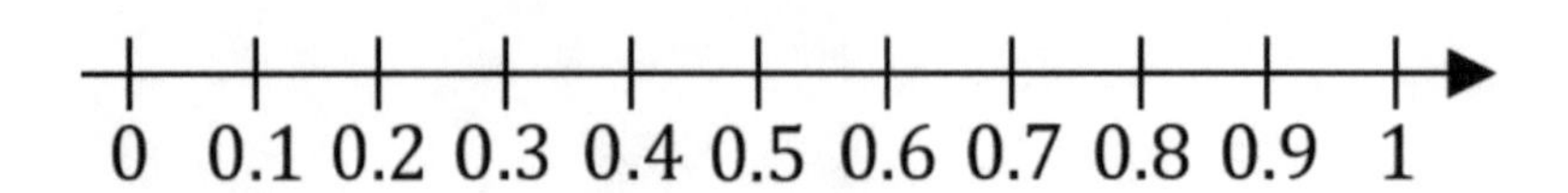

3.7 Ten times greater or smaller

The number 0.3 is 10 times greater than 0.03 because the tenths place is 10 times larger than the hundredths place. Similarly, number 0.03 is 10 times smaller than 0.3.

Examples. **(A)** 0.8 is 10 times larger than 0.08

(B) 0.002 is 10 times smaller than 0.02

Problems. Fill in the blanks.

① ____________ is 10 times larger than 0.007

② ____________ is 10 times smaller than 0.4

③ 0.9 is 10 times larger than ____________

④ 0.05 is 10 times smaller than ____________

⑤ 0.0002 is 10 times ____________ than 0.002

⑥ 6 is 10 times ____________ than 0.6

3.8 Comparing decimals

Recall the less than (<) and greater than (>) symbols. For example, $4 < 7$ and $7 > 4$. To compare two decimals:

- Identify the leftmost nonzero digit of each number.
- If one number begins in a higher place value than the other, that number is larger. For example, $0.1 > 0.01$.
- If the numbers begin in the same place value, compare their leftmost nonzero digits. For example, $0.03 > 0.02$.
- If needed, go to the next digit. For example, $0.56 < 0.59$.
- If necessary, you may add trailing zeroes to a number. For example, 0.4 is the same as 0.40 and 0.400.

Examples. Write $>$, $<$, or $=$ between each pair of numbers.

(A) $0.63 > 0.62$ (B) $0.05 < 0.5$ (C) $0.2 = 0.20$

Problems. Write $>$, $<$, or $=$ between each pair of numbers.

① 0.6 0.4 ② 0.01 0.07 ③ 0.05 0.050

④ 0.3 0.03 ⑤ 0.14 0.13 ⑥ 0.027 0.029

⑦ 0.005 0.04 ⑧ 0.319 0.324 ⑨ 0.634 0.633

3.9 Ordering decimals

To put decimals in order, compare them to determine which is least and which is greatest. For example, out of 0.32, 0.28, and 0.3, the least is 0.28 and the greatest is 0.32. (Note that 0.3 is equivalent to 0.30.) The order is: 0.28, 0.3, and 0.32.

Examples. Order the numbers from least to greatest.

(A) $0.61, 0.59, 0.6 \rightarrow 0.59, 0.6, 0.61$

(B) $0.432, 0.431, 0.434 \rightarrow 0.431, 0.432, 0.434$

Problems. Order the given numbers from least to greatest.

① 0.3, 0.18, 0.24

② 0.07, 0.3, 0.005

③ 0.008, 0.005, 0.009

④ 0.07, 0.3, 0.04

⑤ 0.216, 0.22, 0.21

⑥ 0.75, 0.69, 0.7

⑦ 0.435, 0.429, 0.43

3.10 Rounding decimals

To round a decimal, follow these steps:

- Look at the digit to the right of the digit where you are rounding. For example, to round 0.48 to the tenths place, since the 4 is in the tenths place, look at the 8.
- If the digit one place to the right is 4 or less, the digit where you are rounding doesn't change.
- If the digit one place to the right is 5 or higher, the digit where you are rounding increases by one.
- Either way, remove the digits to the right of the digit where you are rounding.

Examples. (A) Round 0.361 to the nearest tenth.

The 3 is in the tenths place. The 6 is one digit to the right. Since the 6 is 5 or higher, the 3 increases by one. Remove the digits to the right. The final answer is 0.4.

(B) Round 0.472 to the nearest hundredth.

The 7 is in the hundredths place. The 2 is one digit to the right. Since the 2 is 4 or less, the 7 doesn't change. Remove the digits to the right. The final answer is 0.47.

Problems. Round each number as directed.

① Round 0.29 to the nearest tenth.

② Round 0.064 to the nearest hundredth.

③ Round 0.716 to the nearest tenth.

④ Round 0.178 to the nearest hundredth.

⑤ Round 0.1469 to the nearest tenth.

⑥ Round 0.3512 to the nearest hundredth.

⑦ Round 0.5916 to the nearest thousandth.

⑧ Round 0.25 to the nearest tenth.

⑨ Round 0.03844 to the nearest thousandth.

⑩ Round 0.055 to the nearest hundredth.

⑪ Round 0.07 to the nearest tenth.

⑫ Round 0.0099 to the nearest thousandth.

⑬ Round 0.0083 to the nearest hundredth.

⑭ Round 0.00672 to the nearest thousandth.

Multiple Choice Questions

① What is the decimal place of the 4 in 31.498?

(A) units (B) tenths (C) hundredths (D) thousandths

② What is the decimal place of the 8 in 0.7583?

(A) units (B) tenths (C) hundredths (D) thousandths

③ What is the decimal place of the 2 in 6.4271?

(A) units (B) tenths (C) hundredths (D) thousandths

④ Which number equals $50 + 4 + \frac{9}{100}$?

(A) 0.5409 (B) 0.549 (C) 54.09 (D) 54.9 (E) 549

⑤ Which number is thirty-seven hundredths?

(A) 100.37 (B) 0.0037 (C) 0.037 (D) 0.37 (E) 3700

⑥ Which number is four hundred fifteen thousandths?

(A) 0.00415 (B) 0.0415 (C) 0.415 (D) 415.001 (E) 415,000

⑦ Which number is twelve thousandths?

(A) 0.00012 (B) 0.012 (C) 0.12 (D) 1000.12 (E) 12,000

tens	units	.	tenths	hundredths	thousandths
	6	.	4	9	

⑧ Which number appears in the chart above?

(A) 6.49 (B) 6.499 (C) 16.49 (D) 64.9 (E) 649

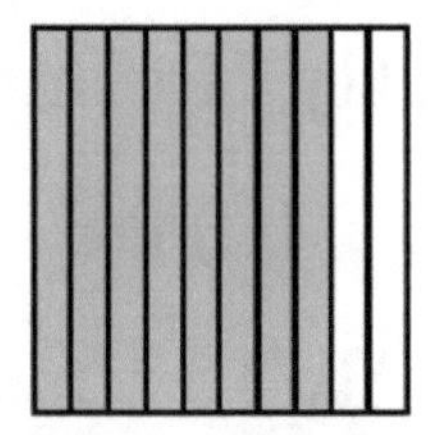

⑨ Which decimal is illustrated above?

(A) 0.4 (B) 0.5 (C) 0.6 (D) 0.7 (E) 0.8

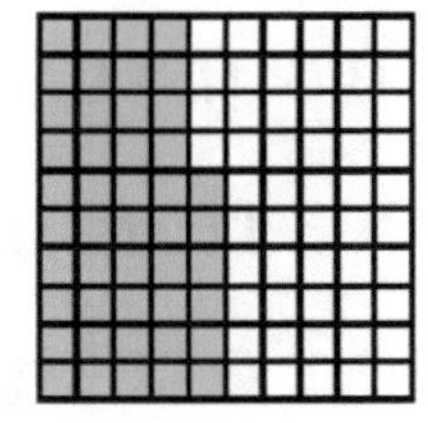

⑩ Which decimal is illustrated above?

(A) 0.45 (B) 0.46 (C) 0.47 (D) 0.56 (E) 0.64

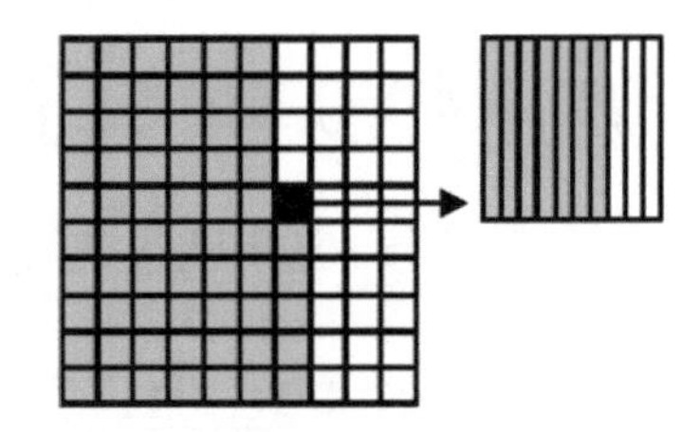

⑪ Which decimal is illustrated above?

(A) 0.656 (B) 0.657 (C) 0.658 (D) 0.756 (E) 0.757

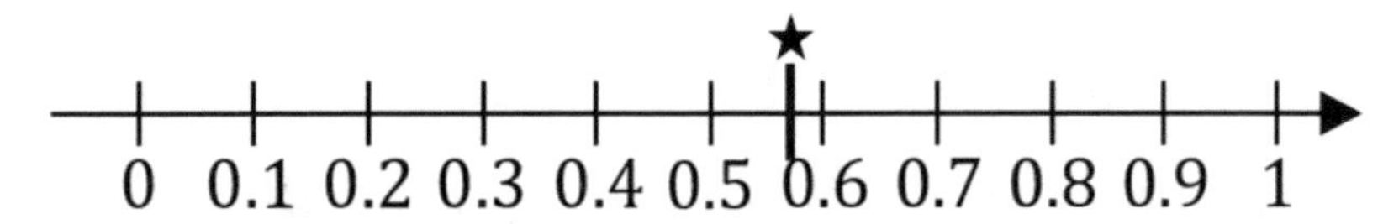

⑫ On the number line above, what is the value of the line that is labeled with a ★?

(A) 0.52 (B) 0.55 (C) 0.58 (D) 0.62 (E) 0.8

⑬ Which number is exactly 10 times **larger** than 0.08?

(A) 0.07 (B) 0.008 (C) 0.8 (D) 0.09 (E) 0.9

⑭ Which number is exactly 10 times **smaller** than 0.03?

(A) 0.02 (B) 0.2 (C) 0.003 (D) 0.3 (E) 0.04

⑮ Which of the following is **NOT** true?

(A) $0.07 < 0.3$ (B) $0.62 > 0.6$ (C) $0.4 < 0.40$ (D) $0.08 > 0.04$

⑯ Order 0.01, 0.2, and 0.003 from least to greatest.

(A) 0.01, 0.2, 0.003 (B) 0.1, 0.003, 0.2 (C) 0.2, 0.01, 0.003

(D) 0.003, 0.2, 0.01 (E) 0.003, 0.01, 0.2

⑰ Order 0.5, 0.49, and 0.51 from least to greatest.

(A) 0.49, 0.51, 0.5 (B) 0.49, 0.5, 0.51 (C) 0.5, 0.49, 0.51

(D) 0.51, 0.49, 0.5 (E) 0.5, 0.51, 0.49

⑱ Round 0.654 to the nearest **tenth**.

(A) 0.6 (B) 0.65 (C) 0.655 (D) 0.66 (E) 0.7

⑲ Round 0.0961 to the nearest **hundredth**.

(A) 0.10 (B) 0.09 (C) 0.096 (D) 0.097 (E) 0.0962

⑳ Round 0.2547 to the nearest **thousandth**.

(A) 0.25 (B) 0.26 (C) 0.254 (D) 0.255 (E) 0.3

-4-

ARITHMETIC WITH DECIMALS

4.1 Add decimals

To add two decimals together, follow these steps:

- Does one number have fewer decimal places than the other? If so, add trailing zeroes. For example, to add $0.32 + 0.587$, first rewrite this as $0.320 + 0.587$.
- Line the numbers up to match the decimal point and the place value of each digit.

$$\begin{array}{r} 0.320 \\ +\,0.587 \\ \hline \end{array}$$

- Add the decimals like you would add whole numbers. Include a decimal point in the same position. As with ordinary addition, it may be necessary to regroup, like the 1 below that comes from $8 + 2 = 10$.

$$\begin{array}{r} {\scriptstyle 1} \\ 0.320 \\ +\,0.587 \\ \hline 0.907 \end{array}$$

Examples. Add the decimals.

(A) $4.6 + 3.75 \rightarrow$

$$\begin{array}{r} {\scriptstyle 1} \\ 4.60 \\ +\,3.75 \\ \hline 8.35 \end{array}$$

(B) $0.248 + 0.36 \rightarrow$

$$\begin{array}{r} {\scriptstyle 1} \\ 0.248 \\ +\,0.360 \\ \hline 0.608 \end{array}$$

Problems. Add the decimals.

① $0.53 + 0.45 =$

② $2.5 + 0.7 =$

③ $0.08 + 0.41 =$

④ $0.8 + 0.41 =$

⑤ $1.75 + 0.5 =$

⑥ $0.374 + 0.562 =$

⑦ $24.7 + 6.93 =$

⑧ $0.679 + 0.54 =$

⑨ $1.18 + 0.343 =$

⑩ $0.06 + 0.193 =$

4.2 Subtract decimals

To subtract two decimals, follow these steps:

- Does one number have fewer decimal places than the other? If so, add trailing zeroes. For example, $2 - 0.75$ becomes $2.00 - 0.75$.
- Line the numbers up to match the decimal point and the place value of each digit.

$$\begin{array}{r} 2.00 \\ -\ 0.75 \\ \hline \end{array}$$

- Subtract the decimals like you would subtract whole numbers. Include a decimal point in the same position. As with ordinary subtraction, it may be necessary to regroup (or "borrow"), like the example below.

$$\begin{array}{r} {\scriptstyle 1\ \ 9\,10} \\ \not{2}.\not{0}\not{0} \\ -\ 0.75 \\ \hline 1.25 \end{array}$$

Examples. Subtract the decimals.

(A) $5.4 - 1.26 \rightarrow$

$$\begin{array}{r} {\scriptstyle 3\,10} \\ 5.\not{4}\not{0} \\ -\ 1.26 \\ \hline 4.14 \end{array}$$

(B) $0.457 - 0.27 \rightarrow$

$$\begin{array}{r} {\scriptstyle 3\,15} \\ 0.\not{4}\not{5}7 \\ -\ 0.270 \\ \hline 0.187 \end{array}$$

Problems. Subtract the decimals.

① $3.8 - 2.3 =$

② $0.7 - 0.18 =$

③ $0.65 - 0.03 =$

④ $0.65 - 0.3 =$

⑤ $0.872 - 0.352 =$

⑥ $9.4 - 4.78 =$

⑦ $7 - 2.6 =$

⑧ $0.851 - 0.36 =$

⑨ $0.082 - 0.0795 =$

⑩ $5.4 - 1.88 =$

4.3 Estimate decimal addition/subtraction

One way to estimate a sum or difference with decimals is to round the decimals (Sec. 3.10). For example, $4.63 + 3.34$ is approximately equal to $5 + 3 = 8$ because 4.63 rounds to 5 and 3.34 rounds to 3. Compare the estimated sum of 8 with the actual value of $4.63 + 3.34 = 7.97$. Be sure to round each number to the **<u>same decimal place</u>**.

Examples. (A) $0.793 + 0.32 \approx 0.8 + 0.3 = 1.1$ (tenths)

(B) $8.15 - 4.96 \approx 8 - 5 = 3$ (rounded to the units place)

(C) $0.0394 + 0.0081 \approx 0.039 + 0.008 = 0.047$ (thousandths)

Problems. Estimate each answer by rounding.

① $7.2 + 5.93 \approx$

② $0.894 - 0.31 \approx$

③ $0.078 + 0.0493 \approx$

④ $24.9 - 6.15 \approx$

⑤ $0.913 + 0.0781 \approx$

⑥ $5.23 - 1.9 \approx$

4.4 Multiply decimals by powers of ten

A power of ten refers to 10, 100, 1000, etc., or to 0.1, 0.01, 0.001, etc. Multiplying by a power of ten has the effect of shifting the decimal point:

- To multiply by 10, shift the decimal point one place to the right. For example, $0.032 \times 10 = 0.32$.
- To multiply by 100, shift the decimal point 2 places to the right. For example, $0.032 \times 100 = 3.2$.
- To multiply by 1000, shift the decimal point 3 places to the right. For example, $0.032 \times 1000 = 32$.
- To multiply by 0.1, shift the decimal point one place to the left. For example, $0.032 \times 0.1 = 0.0032$.
- To multiply by 0.01, shift the decimal point 2 places to the left. For example, $0.032 \times 0.01 = 0.00032$.
- To multiply by 0.001, shift the decimal point 3 places to the left. For example, $0.032 \times 0.001 = 0.000032$.
- If the decimal point shifts beyond the last place, add a zero. For example, $4.7 \times 100 = 470$, $8.5 \times 1000 = 8500$, and $1.6 \times 0.001 = 0.0016$.

Examples. Multiply the numbers.

(A) $0.76 \times 10 = 7.6$ (shift 1 place to the right)

(B) $2.8 \times 0.01 = 0.028$ (shift 2 places to the left)

(C) $4.3 \times 1000 = 4300$ (shift 3 places to the right)

Problems. Multiply the numbers.

① $3.14 \times 10 =$

② $0.0675 \times 100 =$

③ $0.25 \times 0.1 =$

④ $0.048 \times 0.01 =$

⑤ $0.57 \times 1000 =$

⑥ $7.423 \times 0.1 =$

⑦ $2.9 \times 100 =$

⑧ $34.2 \times 0.001 =$

⑨ $1.8795 \times 1000 =$

⑩ $0.707 \times 0.01 =$

⑪ $0.16 \times 0.001 =$

⑫ $0.5 \times 100 =$

⑬ $0.0673 \times 10 =$

⑭ $0.053 \times 0.1 =$

⑮ $2.6 \times 0.01 =$

⑯ $0.79 \times 10 =$

⑰ $6.42 \times 1000 =$

⑱ $8.1 \times 0.001 =$

4.5 Multiply decimals by whole numbers

To multiply a decimal by a whole number, follow these steps:

- Count the number of decimal places initially.
- Multiply the numbers (ignoring the decimal point for the moment) like we did in Sec.'s 1.2-1.5.
- Insert a decimal point so that the answer has the same number of decimal places as in the first step above.
- If your answer has any decimal trailing zeroes, remove them. For example, 0.03600 is equivalent to 0.036.

Examples. Multiply the numbers.

(A)

$$\begin{array}{r} {\scriptstyle 1} \\ 5.4 \\ \times\ 4 \\ \hline 21.6 \end{array}$$

(B)

$$\begin{array}{r} {\scriptstyle 3\ 2} \\ 3.64 \\ \times\ 5 \\ \hline 18.20 \end{array}$$

$= 18.2$

(C)

$$\begin{array}{rl} {\scriptstyle 2} & \\ {\scriptstyle 1} & \\ 3.6 & \leftarrow \text{one decimal place} \\ \times\ 4\ 2 & \leftarrow \text{zero decimals} \\ \hline 7.2 & \\ 144.0 & \\ \hline 151.2 & \leftarrow \text{one decimal place} \end{array}$$

Problems. Multiply the numbers.

①

$$\begin{array}{r} 0.47 \\ \times\ 6 \\ \hline \end{array}$$

②

$$\begin{array}{r} 8.59 \\ \times\ 8 \\ \hline \end{array}$$

③

$$\begin{array}{r} 8.6 \\ \times\ 7 \\ \hline \end{array}$$

④

$$\begin{array}{r} 0.57 \\ \times\ 12 \\ \hline \end{array}$$

⑤

$$\begin{array}{r} 7.4 \\ \times\ 3\ 6 \\ \hline \end{array}$$

⑥

$$\begin{array}{r} 0.639 \\ \times\ 4 \\ \hline \end{array}$$

⑦

$$\begin{array}{r} 47.3 \\ \times\ 9 \\ \hline \end{array}$$

⑧

$$\begin{array}{r} 2.88 \\ \times\ 68 \\ \hline \end{array}$$

An alternative way to multiply a one-digit whole number by a decimal is to use the distributive property (Sec. 1.13): $a \times (b + c) = a \times b + a \times c$. For example, for 7×2.5, we can identify $a = 7$, $b = 2$, and $c = 5$ to write:

$$7 \times 2.5 = 7 \times (2 + 0.5) = 7 \times 2 + 7 \times 0.5 = 14 + 3.5 = 17.5$$

Examples. Apply the distributive property.

(D) $3 \times 0.64 = 3 \times (0.6 + 0.04) = 3 \times 0.6 + 3 \times 0.04 = 1.8 + 0.12 = 1.92$

(E) $5 \times 2.7 = 5 \times (2 + 0.7) = 5 \times 2 + 5 \times 0.7 = 10 + 3.5 = 13.5$

Problems. Apply the distributive property.

① $4 \times 7.3 =$

② $6 \times 0.15 =$

③ $3 \times 8.7 =$

④ $8 \times 0.049 =$

⑤ $9 \times 0.58 =$

⑥ $2 \times 0.097 =$

⑦ $5 \times 0.888 =$

⑧ $7 \times 6.94 =$

4.6 Multiply one-digit decimals

To multiply two decimals where each decimal only has one nonzero digit, follow these steps:

- Count the total number of decimal places initially.
- Multiply the numbers (ignoring the decimal point for the moment).
- Insert a decimal point so that the answer has the same number of decimal places as in the first step above.
- If your answer has any decimal trailing zeroes, remove them. For example, 0.80 is equivalent to 0.8.

Examples. Multiply the numbers.

(A) $0.3 \times 0.2 = 0.06$ (the answer has 2 decimal places)

(B) $0.4 \times 0.06 = 0.024$ (the answer has 3 decimal places)

(C) $0.5 \times 0.4 = 0.20 = 0.2$ (remove the trailing decimal zero)

Problems. Multiply the numbers.

① $0.8 \times 0.3 =$

② $0.7 \times 0.5 =$

③ $0.4 \times 0.04 =$

④ $0.09 \times 0.06 =$

⑤ $0.8 \times 0.8 =$

⑥ $0.06 \times 0.03 =$

⑦ $0.7 \times 0.06 =$

⑧ $0.5 \times 0.2 =$

⑨ $0.08 \times 0.04 =$

⑩ $0.1 \times 0.01 =$

⑪ $0.6 \times 0.006 =$

⑫ $0.09 \times 0.08 =$

⑬ $0.8 \times 0.5 =$

⑭ $0.053 \times 0.1 =$

⑮ $0.09 \times 0.7 =$

⑯ $0.009 \times 0.3 =$

⑰ $0.002 \times 0.004 =$

⑱ $0.09 \times 0.009 =$

4.7 Multiply decimals in expanded form

Recall the expanded form of a decimal from Sec. 3.2. For example, 4.8 may be expanded as $4 + \frac{8}{10}$. We may also expand 4.8 using a decimal rather than a fraction as $4 + 0.8$. The expanded form with a decimal may be used to multiply two-digit decimals together.

Consider a rectangle that is 2.4 m long and 1.5 m wide. We can find the area of the rectangle by multiplying the length and width. If we write these numbers in expanded form, the area is $2.4 \times 1.5 = (2 + 0.4) \times (1 + 0.5)$. This effectively divides the rectangle into four parts, as shown below. (The diagram isn't drawn to scale, but helps to visualize how the rectangle is divided into four parts).

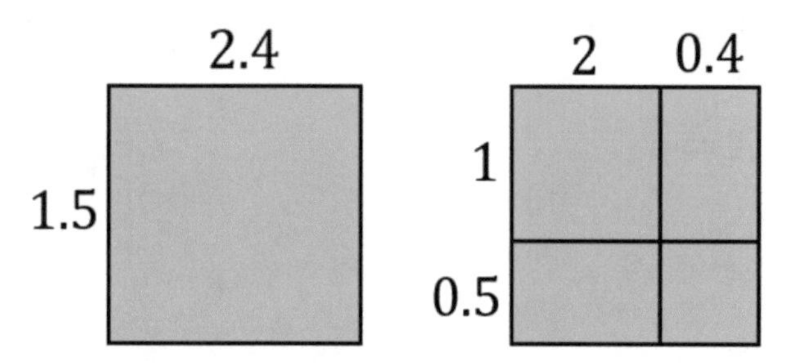

The diagram shows that we can find the area of the big rectangle by adding together the areas of the four parts:

$$2.4 \times 1.5 = (2 + 0.4) \times (1 + 0.5)$$
$$= 2 \times 1 + 2 \times 0.5 + 0.4 \times 1 + 0.4 \times 0.5$$
$$= 2 + 1 + 0.4 + 0.20 = 3.60 = 3.6$$

Example. Multiply the decimals in expanded form.

(A) $4.5 \times 3.6 =$

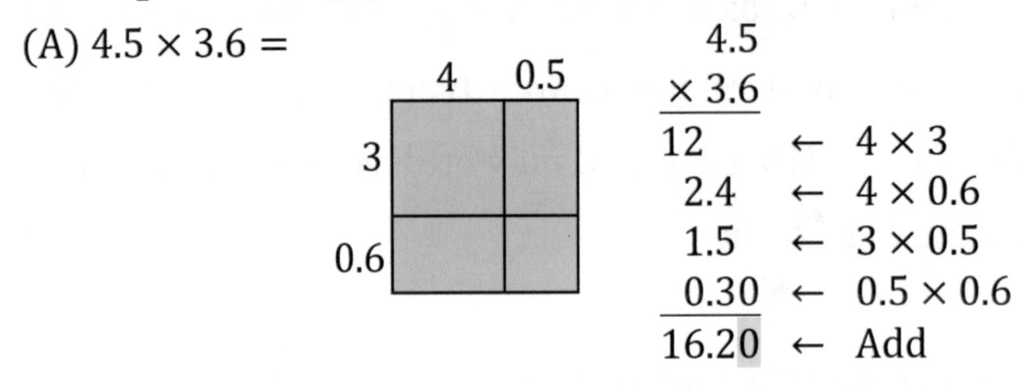

- We expanded $4.5 \times 3.6 = (4 + 0.5) \times (3 + 0.6)$.
- We found the area of each rectangle: $4 \times 3 = 12$, $4 \times 0.6 = 2.4$, $3 \times 0.5 = 1.5$, and $0.5 \times 0.6 = 0.30$.
- We lined these four values up at their decimal points and added: $12 + 2.4 + 1.5 + 0.30 = 16.20$.
- Remove the trailing zero: 16.2 is the final answer.

Problems. Multiply the decimals in expanded form.

① $7.2 \times 4.3 =$

② $9.6 \times 7.8 =$

③ $8.4 \times 0.53 =$

④ $0.42 \times 0.039 =$

⑤ $0.38 \times 0.27 =$

4.8 Visualize decimal multiplication

In Sec. 4.7, we saw how to picture decimal multiplication as the area of a rectangle. In this section, we will visualize decimal multiplication using a decimal square. A decimal square is a 10×10 grid. The complete grid represents one unit. Each tiny square represents one hundredth (0.01).

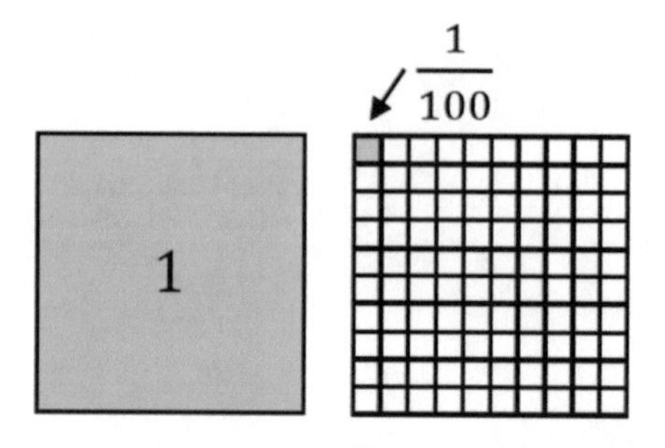

To multiply two decimals, shade a rectangle in the decimal square using the decimals for the length and width. Find the area of the rectangle to determine the product. The example below illustrates 0.4×0.3. Since the shaded rectangle has 12 tiny squares and each square is $\frac{1}{100}$, the product is $0.4 \times 0.3 = \frac{12}{100} = 0.12$.

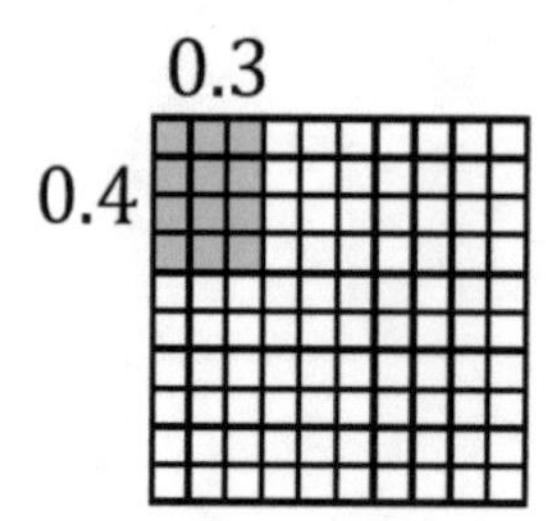

Note: Some texts shade two rectangles with two different colors. They first shade every column of the width (0.3 in the previous example) and then shade the rows a different color, like the example below. The idea behind this method is to first find 0.3 and then find 0.4 of 0.3 to get 0.12. The two methods are equivalent. Our method (on the previous page) just shades one rectangle, focusing on the answer.

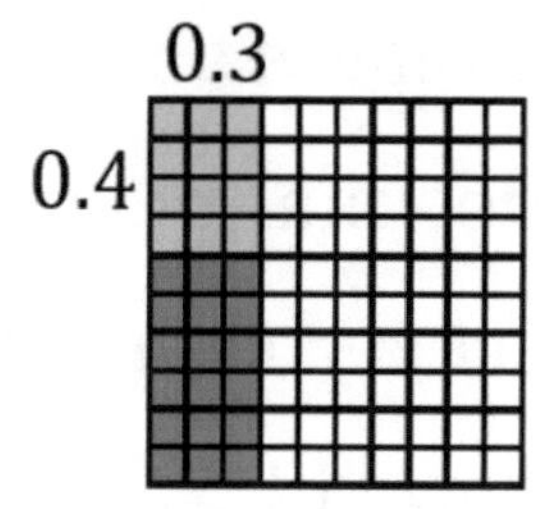

Example. (A) Draw 0.6×0.4 and determine the answer.

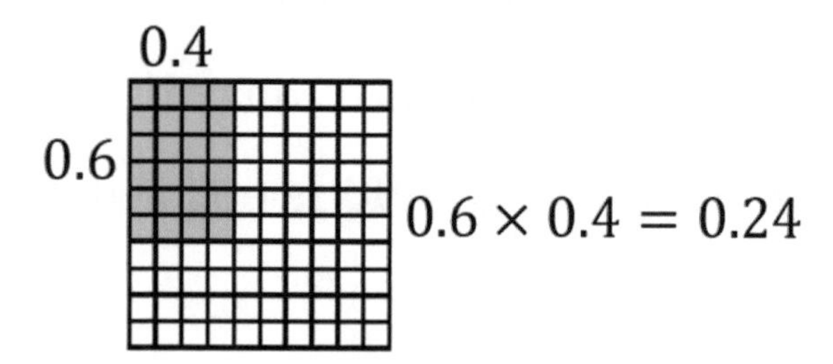

Problems. Draw each problem and determine the answer.

① $0.7 \times 0.4 =$

② $0.9 \times 0.6 =$

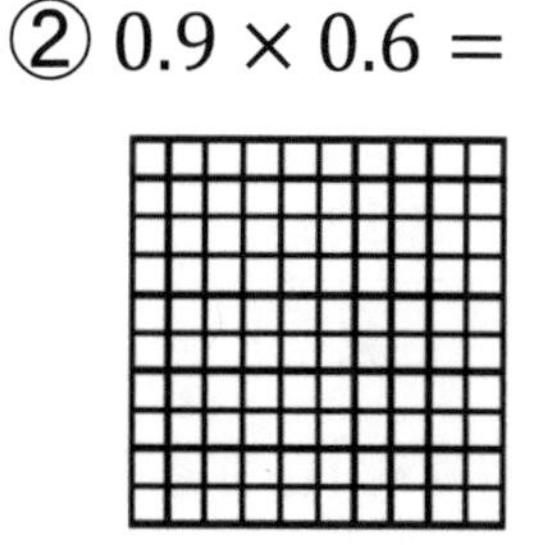

4.9 Multiply decimals together

To multiply decimals together, follow these steps:

- Count the total number of decimal places initially.
- Multiply the numbers (ignoring the decimal point for the moment) like we did in Sec.'s **1.4-1.5**.
- Insert a decimal point so that the answer has the same number of decimal places as in the first step above.
- If you need more decimal places than there are digits in the answer, insert zeroes at the left. For example, to express 32 with a total of 5 decimal places, first write 00032 and then add the decimal point: 0.00032.
- If your answer has any decimal trailing zeroes, remove them. For example, 0.07600 is equivalent to 0.076.

Example. Multiply the numbers.

(A)

$$
\begin{array}{lrcrcl}
 & & & {}^{2}_{1} & & \\
 & 0.63 & & 63 & & \\
 & \underline{\times\, 0.74} & \rightarrow & \underline{\times\, 74} & & \\
 & & & 252 & & \\
2 + 2 = 4 & & & \underline{4410} & & 4 \text{ decimal places} \\
\text{decimal places} & & & 4662 & \rightarrow & 0.4662
\end{array}
$$

Since 0.63×0.74 has a total of 4 decimal places, the answer is 0.4662.

Problems. Multiply the numbers.

①

$$\begin{array}{r} 0.84 \\ \times\ 0.46 \\ \hline \end{array}$$

②

$$\begin{array}{r} 5.7 \\ \times\ 0.34 \\ \hline \end{array}$$

③

$$\begin{array}{r} 6.3 \\ \times\ 2.6 \\ \hline \end{array}$$

④

$$\begin{array}{r} 0.77 \\ \times\ 9.2 \\ \hline \end{array}$$

⑤

$$\begin{array}{r} 0.94 \\ \times\ 0.78 \\ \hline \end{array}$$

⑥

$$\begin{array}{r} 0.059 \\ \times\ 0.48 \\ \hline \end{array}$$

⑦

$$\begin{array}{r} 2.47 \\ \times\ 0.38 \\ \hline \end{array}$$

⑧

$$\begin{array}{r} 0.479 \\ \times\ 0.62 \\ \hline \end{array}$$

⑨

$$\begin{array}{r} 0.788 \\ \times\ 0.95 \\ \hline \end{array}$$

⑩

$$\begin{array}{r} 8.36 \\ \times\ 6.7 \\ \hline \end{array}$$

⑪

$$\begin{array}{r} 0.693 \\ \times\ 0.84 \\ \hline \end{array}$$

⑫

$$\begin{array}{r} 0.0924 \\ \times\ 0.73 \\ \hline \end{array}$$

4.10 Estimate decimal multiplication

Decimal multiplication can be estimated by rounding each decimal (similar to Sec.'s 1.7-1.8). For example, 0.21×0.39 is approximately equal to $0.2 \times 0.4 = 0.08$.

Examples. Estimate the answers by rounding.

(A) $4.92 \times 0.31 \approx 5 \times 0.3 = 1.5$

(B) $0.61 \times 0.194 \approx 0.6 \times 0.2 = 0.12$

Problems. Estimate the answers by rounding.

① $0.713 \times 0.687 \approx$

② $1.89 \times 0.72 \approx$

③ $0.794 \times 0.031 \approx$

④ $0.406 \times 0.49 \approx$

⑤ $7.88 \times 5.96 \approx$

⑥ $0.061 \times 0.0807 \approx$

⑦ $19.8 \times 0.42 \approx$

⑧ $0.098 \times 0.513 \approx$

4.11 Divide decimals by powers of ten

A power of ten refers to 10, 100, 1000, etc., or to 0.1, 0.01, 0.001, etc. Dividing by a power of ten has the effect of shifting the decimal point:

- To divide by 10, shift the decimal point one place to the left. For example, $0.032 \div 10 = 0.0032$.
- To divide by 100, shift the decimal point 2 places to the left. For example, $0.032 \div 100 = 0.00032$.
- To divide by 1000, shift the decimal point 3 places to the left. For example, $0.032 \div 1000 = 0.000032$.
- To divide by 0.1, shift the decimal point one place to the right. For example, $0.032 \div 0.1 = 0.32$.
- To divide by 0.01, shift the decimal point 2 places to the right. For example, $0.032 \div 0.01 = 3.2$.
- To divide by 0.001, shift the decimal point 3 places to the right. For example, $0.032 \div 0.001 = 32$.
- If the decimal point shifts beyond the last place, add a zero. For example, $5.2 \div 100 = 0.052$ and $0.2 \div 0.01 = 20$.

Note that dividing by a power of ten has the opposite effect of multiplying by a power of ten (Sec. 4.4).

Examples. Divide the numbers.

(A) $0.49 \div 10 = 0.049$ (shift 1 place to the left)

(B) $0.58 \div 100 = 0.0058$ (shift 2 places to the left)

(C) $3.2 \div 0.001 = 3200$ (shift 3 places to the right)

Problems. Divide the numbers.

① $0.328 \div 100 =$

② $9.2 \div 10 =$

③ $0.78 \div 0.1 =$

④ $0.63 \div 0.01 =$

⑤ $4.5 \div 1000 =$

⑥ $0.8 \div 0.001 =$

⑦ $0.397 \div 10 =$

⑧ $17 \div 0.1 =$

⑨ $6.249 \div 100 =$

⑩ $0.0381 \div 0.001 =$

⑪ $0.8324 \div 0.01 =$

⑫ $0.736 \div 1000 =$

⑬ $0.0244 \div 10 =$

⑭ $1.09 \div 0.001 =$

⑮ $9.0503 \div 0.01 =$

⑯ $0.505 \div 100 =$

⑰ $0.01 \div 1000 =$

⑱ $0.00081 \div 0.1 =$

4.12 Divide decimals by whole numbers

To divide a decimal by a whole number, follow these steps:

- Divide the numbers like we did in Sec.'s 2.2 and 2.4-2.5, preserving the location of the decimal point.
- If your answer has any decimal trailing zeroes, remove them. For example, 2.70 is equivalent to 2.7.
- You may need to add trailing zeroes to the dividend. See Example D.
- It's possible for the answer to have more decimal places than the dividend. See Examples B and D.

Examples. Divide the numbers.

(A)
$$\begin{array}{r} 2.3 \\ 4\overline{)9.2} \\ \underline{8.0} \\ 1.2 \end{array}$$

(B)
$$\begin{array}{r} 0.125 \\ 6\overline{)0.75} \\ \underline{0.6} \\ 0.15 \\ \underline{0.12} \\ 0.03 \end{array}$$

(C)
$$\begin{array}{r} 0.25 \\ 14\overline{)3.5} \\ \underline{2.8} \\ 0.7 \end{array}$$

(D)
$$\begin{array}{r} 0.075 \\ 4\overline{)0.30} \\ \underline{0.28} \\ 0.02 \end{array}$$

- In (D), we added a trailing zero to 0.3 to subtract 0.28.
- In (B), note that $0.02 \times 6 = 0.12$ and $0.005 \times 6 = 0.03$.
- Check the answers by multiplying: $4 \times 2.3 = 9.2$, $6 \times 0.125 = 0.75$, $14 \times 0.25 = 3.5$, and $4 \times 0.075 = 0.3$.

Problems. Divide the numbers.

① $6\overline{)0.78}$

② $9\overline{)34.56}$

③ $5\overline{)2.22}$

④ $8\overline{)0.5}$

⑤ $3\overline{)0.0228}$

⑥ $7\overline{)0.6048}$

⑦

$16\overline{)7.52}$

⑧

$52\overline{)0.624}$

⑨

$83\overline{)5.395}$

⑩

$44\overline{)0.033}$

⑪

$75\overline{)0.3}$

⑫

$94\overline{)35.25}$

4.13 Visualize decimal division

We can draw pictures to visualize a decimal divided by a whole number. One way to do this is with base-ten blocks:

- A large square represents a single unit. The square can be divided into 100 tiny squares.
- A long strip represents one tenth (0.1). The strip can be divided into 10 tiny squares.
- A tiny square represents one hundredth (0.01).

(Note how these base-ten blocks are different compared to the base-ten blocks for whole number division in Sec. 2.3.)

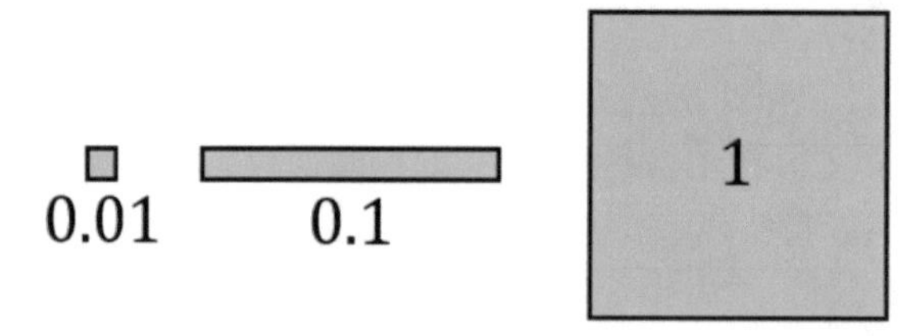

We can use base-ten blocks to draw a decimal divided by a whole number following these steps:

- First draw squares, strips, and tiny squares to add up to the dividend. For example, for $1.32 \div 4$, the dividend (1.32) is 1 square, 3 strips, and 2 tiny squares.
- Rearrange the squares, strips, and tiny squares into as many groups as the divisor. For $1.32 \div 4$, make 4 groups (since the divisor equals 4).

- The groups need to have equal value.
- It is usually necessary to regroup. A large square can be redrawn as 10 strips. A strip can be redrawn as 10 tiny squares.
- Once the groups have equal value and all of the squares, strips, and tiny squares have been used, determine the value of each group. This is the answer to the problem.

An example of how to draw $1.32 \div 4$ is illustrated below.

First draw the dividend.

$1 + 0.3 + 0.02 = 1.32$

We regrouped the square into 10 strips, and we regrouped one strip into 10 tiny squares.

We rearranged the strips and tiny squares into 4 groups. Each group is circled.

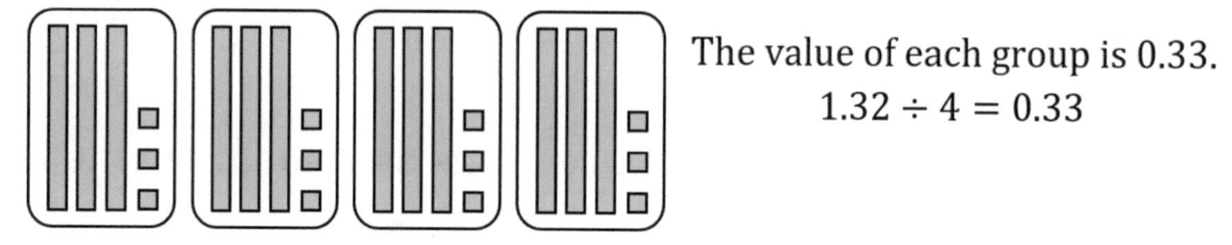

The value of each group is 0.33.
1.32 ÷ 4 = 0.33

We divided 1.32 (one large square plus 3 strips plus 2 tiny squares) into 4 groups of 0.33 (3 strips plus 3 tiny squares). The answer to $1.32 \div 4$ equals 0.33.

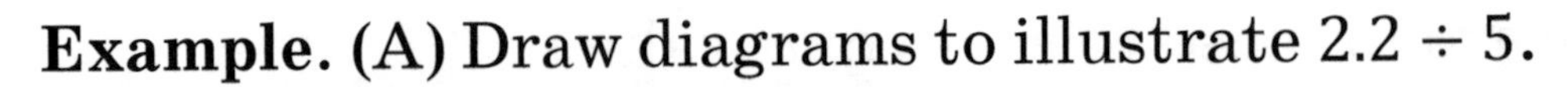

Example. (A) Draw diagrams to illustrate $2.2 \div 5$.

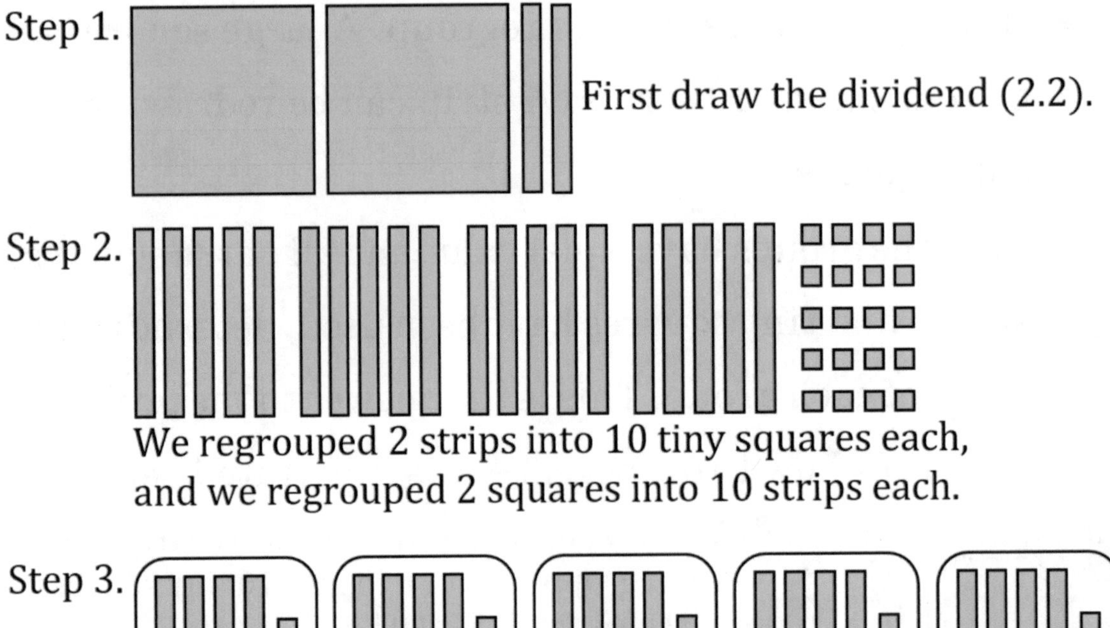

Step 1. First draw the dividend (2.2).

Step 2. We regrouped 2 strips into 10 tiny squares each, and we regrouped 2 squares into 10 strips each.

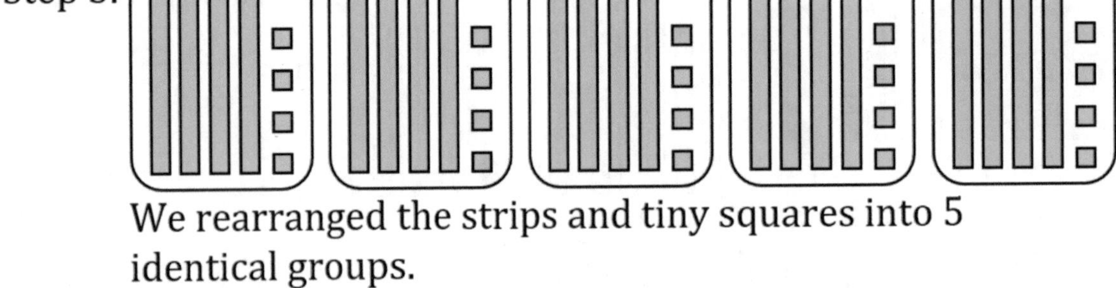

Step 3. We rearranged the strips and tiny squares into 5 identical groups.

Step 4. Each group has a value of 0.44. The answer to $2.2 \div 5$ is 0.44.

- The dividend is 2.2. We drew this as 2 squares plus 2 strips.
- We regrouped each strip into 10 tiny squares and each large square into 10 strips.
- We rearranged the strips and tiny squares into 5 groups of 4 strips plus 4 tiny squares. The answer is 0.44.

Problems. Draw diagrams to solve each problem visually.

① $0.84 \div 6 =$

(A) Draw the dividend.

(B) Redraw the picture with regrouping.

(C) Draw the final picture showing the answer.

(D) According to the final picture, what is $0.84 \div 6$?

② $2.16 \div 8 =$

(A) Draw the dividend.

(B) Redraw the picture with regrouping.

(C) Draw the final picture showing the answer.

(D) According to the final picture, what is $2.16 \div 8$?

4.14 Estimate decimal division

One way to estimate a decimal divided by a whole number is to round the decimal to a number that the whole number divides into evenly. For example, $0.31 \div 4$ is approximately equal to $0.32 \div 4 = 0.08$. Check the answer by multiplying: $4 \times 0.08 = 0.32$.

Examples. Estimate the answers by rounding.

(A) $1.13 \div 3 \approx 1.2 \div 3 = 0.4$ Check: $3 \times 0.4 = 1.2$

(B) $0.361 \div 5 \approx 0.35 \div 5 = 0.07$ Check: $5 \times 0.07 = 0.35$

Problems. Estimate the answers by rounding.

① $4.092 \div 6 \approx$

② $0.394 \div 8 \approx$

③ $12.21 \div 5 \approx$

④ $0.0193 \div 9 \approx$

⑤ $3.498 \div 4 \approx$

⑥ $5.555 \div 3 \approx$

⑦ $0.5 \div 7 \approx$

4.15 Inverse operations

Arithmetic operations that are basically the opposite of one another are called **inverse** operations. Two common pairs of inverse operations are:

- addition and subtraction. Every subtraction problem can be rewritten using addition. For example, $11 - 5 = 6$ can be rewritten as $6 + 5 = 11$. Subtracting 5 is the opposite of adding 5. If you subtract 5 and then add 5, you end up where you started.
- multiplication and division. Every division problem can be rewritten using multiplication. For example, $12 \div 3 = 4$ can be rewritten as $4 \times 3 = 12$. Dividing by 3 is the opposite of multiplying by 3. If you divide by 3 and then multiply by 3, you end up where you started.

You can use the inverse operation to check the answer to an arithmetic problem. For example, if you subtract decimals to get $0.8 - 0.24 = 0.56$, you can check this using addition: $0.56 + 0.24 = 0.8$. As another example, if you divide decimals to get $0.28 \div 0.4 = 7$, you can check this using multiplication: $7 \times 0.4 = 0.28$.

Examples. Use the inverse operation to check the answer.

(A) $0.6 - 0.36 \approx 0.24$ Check: $0.24 + 0.36 = 0.60 = 0.6$

(B) $2.1 \div 7 \approx 0.3$ Check: $7 \times 0.3 = 2.1$

Problems. Use the inverse operation to check the answer.

① $0.8 + 0.6 =$

② $0.16 - 0.07 =$

③ $4 \times 0.9 =$

④ $0.42 \div 7 =$

⑤ $1.2 - 0.47 =$

⑥ $0.62 \times 5 =$

⑦ $0.056 + 0.0056 =$

⑧ $3.2 \div 8 =$

⑨ $3 - 0.84 =$

⑩ $0.054 \div 9 =$

4.16 Word problems with decimals

Reason out which arithmetic operation is involved, or if the solution requires combining operations together. It may be helpful to review Sec.'s 1.9 and 2.10.

Examples. (A) A student lives 7.5 miles from a library. The student rides a bicycle 4.8 miles towards the library. How much farther does the student need to travel?

Subtract the values to find the difference:

$$\begin{array}{r} \scriptstyle 6\ 15 \\ \not{7}.\not{5} \\ -\,4.8 \\ \hline 2.7 \end{array}$$

The student is 2.7 miles from the library.

(B) An empty backpack weighs 4.27 pounds. A student puts 4 notebooks inside of the backpack. Each notebook weighs 0.76 pounds. How much does the backpack weigh now?

Multiply 4 by 0.76 and add the product to 4.27.

$$\begin{array}{r} \scriptstyle 2 \\ 0.76 \\ \times\,4 \\ \hline 3.04 \end{array} \qquad \begin{array}{r} \scriptstyle 1 \\ 3.04 \\ +\,4.27 \\ \hline 7.31 \end{array}$$

The backpack now weighs 7.31 pounds.

Problems. Solve each word problem.

① One rock weighs 3.72 pounds. Another rock weighs 2.63 pounds. What is the combined weight of the rocks?

② If ten identical packets weigh 3.42 ounces, how much does one packet weigh?

③ Regina ran a race in 7.294 seconds. Sydney finished the race 0.368 seconds before Regina. Determine the amount of time it took for Sydney to run the race.

④ A rectangle has a length of 8.6 inches and a width of 4.7 inches. What is the area of the rectangle?

⑤ A customer purchases eight apples and six oranges. Each apple costs $0.49 and each orange costs $0.39. What is the total cost? (There is no sales tax.)

⑥ A glass contains 0.78 liters of water, one-third as much ice as water (by volume), and one-sixth as much lemon juice as water (by volume). What is the combined volume?

4.17 Decimal calculations

It may help to review Sec.'s 1.10-1.12 and 2.11-2.13. Recall that we do math in parentheses first and that we perform multiplication and division from left to right before doing addition and subtraction from left to right.

Examples. **(A)** $0.9 - 0.2 \times 3 = 0.9 - 0.6 = 0.3$ (multiply first)

(B) $(0.4 + 0.3) \div 10 = 0.7 \div 10 = 0.07$ (parentheses first)

Problems. Determine the answer to each problem.

① $0.2 + 0.8 \times 3 =$

② $1.8 \div 3 + 0.4 =$

③ $0.3 \times 0.2 - 0.041 =$

④ $0.9 - 0.6 \div 4 =$

⑤ $(0.36 + 0.54) \times 0.8 =$

⑥ $2.8 \div (0.073 - 0.063) =$

⑦ $(0.87 - 0.77) \times (0.38 + 0.46) =$

⑧ $(2.5 + 1.7) \div (1 - 0.4) =$

Multiple Choice Questions

① What is $0.479 + 0.363$?

(A) 0.732 (B) 0.742 (C) 0.822 (D) 0.832 (E) 0.842

② What is $7 - 0.24$?

(A) 0.54 (B) 0.56 (C) 5.06 (D) 6.76 (E) 6.86

③ Estimate $0.09106 + 0.06893$.

(A) 0.09 (B) 0.15 (C) 0.16 (D) 1.5 (E) 1.6

④ Estimate $0.2934 - 0.1096$.

(A) 0.01 (B) 0.02 (C) 0.03 (D) 0.1 (E) 0.2

⑤ What is 0.43×100?

(A) 0.0043 (B) 0.043 (C) 4.3 (D) 43 (E) 430

⑥ What is 0.0579×0.1?

(A) 0.00579 (B) 0.579 (C) 5.79 (D) 57.9 (E) 579

⑦ What is 6.39×7?

(A) 42.73 (B) 42.73 (C) 44.23 (D) 44.73 (E) 49.73

⑧ What is 0.06×0.005?

(A) 0.00003 (B) 0.0003 (C) 0.003 (D) 0.03 (E) 0.3

⑨ Which number is equivalent to $(7 + 0.4) \times (4 + 0.9)$?

(A) 28.36 (B) 29.96 (C) 31.6 (D) 34.66 (E) 36.26

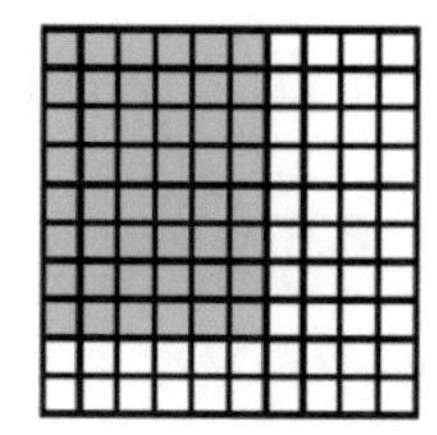

⑩ The diagram above illustrates ____________.

(A) 0.08×0.06 (B) 0.8×0.06 (C) 0.8×0.6 (D) 8×0.6 (E) 0.8×6

⑪ What is 7.9×8.6?

(A) 56.54 (B) 60.8 (C) 60.94 (D) 63.2 (E) 67.94

⑫ Estimate 0.5893×0.3127.

(A) 0.15 (B) 0.18 (C) 0.2 (D) 1.5 (E) 1.8

⑬ What is $79.2 \div 1000$?

(A) 0.000792 (B) 0.0792 (C) 0.792 (D) 7920 (E) 79,200

⑭ What is $0.0627 \div 0.01$?

(A) 0.0000627 (B) 0.000627 (C) 0.00627 (D) 0.627 (E) 6.27

⑮ What is $0.15 \div 4$?

(A) 0.033 (B) 0.0375 (C) 0.33 (D) 0.375 (E) 0.6

⑯ Which operation is the inverse of multiplication?

(A) addition (B) subtraction (C) division (D) estimation

⑰ A boy has 7 small blocks and 4 large blocks. Each small block weighs 6.3 ounces and each large block weighs 18.6 ounces. What is the combined weight of all of the blocks?

(A) 110.5 oz (B) 112.5 oz (C) 114.5 oz (D) 116.5 oz (E) 118.5 oz

⑱ Three friends earned \$5 for washing a car. They decide to split the money evenly among them. How much money should each friend receive?

(A) \$1.33 and there will be 1 penny left over

(B) \$1.34 and there will be 2 pennies left over

(C) \$1.66 and there will be 2 pennies left over

(D) \$1.67 and there will be 1 penny left over

(E) exactly \$1.70

⑲ What is $0.7 + 0.3 \times 6$?

(A) 0.6 (B) 0.88 (C) 0.9 (D) 2.5 (E) 6

⑳ What is $(0.94 - 0.08) \times 0.2$?

(A) 0.028 (B) 0.172 (C) 0.28 (D) 0.66 (E) 1.72

-5-

ARITHMETIC WITH FRACTIONS

5.1 Reducing fractions

If the numerator and denominator share a common factor, the fraction can be **reduced** by dividing the numerator and denominator each by the **greatest common factor** (GCF). It may help to review Sec. 2.15. For example, $\frac{6}{9}$ can be reduced because 6 and 9 are each evenly divisible by 3. The greatest common factor (GCF) of 6 and 9 is 3. Divide 6 and 9 each by 3 to reduce the fraction:

$$\frac{6}{9} = \frac{6 \div 3}{9 \div 3} = \frac{2}{3}$$

Examples. Reduce each fraction to its simplest form.

(A) $\frac{12}{16} = \frac{12 \div 4}{16 \div 4} = \frac{3}{4}$

(B) $\frac{25}{10} = \frac{25 \div 5}{10 \div 5} = \frac{5}{2}$

Problems. Reduce each fraction to its simplest form.

① $\frac{7}{14} =$

② $\frac{12}{8} =$

③ $\frac{12}{18} =$

④ $\frac{15}{25} =$

⑤ $\frac{27}{36} =$

⑥ $\frac{16}{20} =$

⑦ $\frac{15}{9} =$

⑧ $\frac{24}{30} =$

⑨ $\frac{28}{16} =$

⑩ $\frac{56}{35} =$

⑪ $\frac{18}{54} =$

⑫ $\frac{32}{72} =$

5.2 Mixed numbers

If a fraction is larger than one, there are two common ways that the fraction may be written:

- A **mixed number** adds a whole number to a fraction. For example, the mixed number $2\frac{3}{4}$ means 2 and $\frac{3}{4}$. The 2 and $\frac{3}{4}$ are added together in $2\frac{3}{4}$. The mixed number $2\frac{3}{4}$ is larger than 2 by $\frac{3}{4}$.
- An **improper** fraction has a numerator that is greater than the denominator. For example, $\frac{11}{4}$ is an improper fraction because 11 is greater than 4.

To convert a mixed number of the form $a\frac{b}{c}$ into an improper fraction, use the formula $a\frac{b}{c} = \frac{a \times c + b}{c}$. For example, compare $2\frac{3}{4}$ with $a\frac{b}{c}$ to see that $a = 2$, $b = 3$, and $c = 4$:

$$2\frac{3}{4} = \frac{a \times c + b}{c} = \frac{2 \times 4 + 3}{4} = \frac{8 + 3}{4} = \frac{11}{4}$$

Examples. Make an equivalent improper fraction.

(A) $4\frac{2}{5} = \frac{4 \times 5 + 2}{5} = \frac{20 + 2}{5} = \frac{22}{5}$ (B) $3\frac{1}{4} = \frac{3 \times 4 + 1}{4} = \frac{12 + 1}{4} = \frac{13}{4}$

Problems. Make an equivalent improper fraction.

① $3\frac{4}{5} =$

② $6\frac{5}{7} =$

③ $4\frac{1}{3} =$

④ $7\frac{3}{4} =$

⑤ $1\frac{2}{3} =$

⑥ $9\frac{7}{8} =$

⑦ $5\frac{1}{6} =$

⑧ $8\frac{3}{8} =$

To convert an improper fraction of the form $\frac{a}{b}$ into a mixed number, follow these steps:

- Perform the division $a \div b$ to get a whole number plus a remainder. For example, $19 \div 5 = 3\ R4$ (three with a remainder of four) because $3 \times 5 = 15$ and $19 - 15 = 4$.
- The mixed number equals the whole number plus the remainder divided by the original denominator.

Examples. Make an equivalent mixed number.

(A) $\frac{11}{4} = 11 \div 4 = 2\ R3 = 2\frac{3}{4}$ since $2 \times 4 = 8$ and $11 - 8 = 3$

(B) $\frac{16}{3} = 16 \div 3 = 5\ R1 = 5\frac{1}{3}$ since $5 \times 3 = 15$ and $16 - 15 = 1$

These solutions may alternatively be expressed as:

(A) $\frac{11}{4} = \frac{8+3}{4} = \frac{8}{4} + \frac{3}{4} = 2\frac{3}{4}$ (B) $\frac{16}{3} = \frac{15+1}{3} = \frac{15}{3} + \frac{1}{3} = 5\frac{1}{3}$

Problems. Make an equivalent mixed number.

① $\frac{9}{2} =$

② $\frac{13}{5} =$

③ $\frac{29}{6} =$

④ $\frac{20}{3} =$

⑤ $\frac{37}{8} =$

⑥ $\frac{15}{2} =$

⑦ $\frac{18}{7} =$

⑧ $\frac{25}{3} =$

⑨ $\frac{34}{9} =$

⑩ $\frac{39}{4} =$

⑪ $\frac{59}{8} =$

⑫ $\frac{49}{6} =$

5.3 Lowest common denominator

The **lowest common denominator** (LCD) is the same as the least common multiple (LCM) of the denominators. It may help to review Sec. 2.16. As we will explore in Sec.'s 5.4-5.5, it is useful to express two or more given fractions with their LCD when adding, subtracting, or comparing two or more fractions. To find the LCD, follow these steps:

- First determine the LCM (like we did in Sec. 2.16).
- Multiply both the numerator and denominator of each fraction by the factor needed to make the LCD.

Examples. Express each set of fractions with their LCD.

(A) Given $\frac{2}{3}$ and $\frac{1}{4}$, the LCM of 3 and 4 equals 12.

What do you need to multiply $\frac{2}{3}$ by in order to make a denominator of 12? The answer is $\frac{4}{4}$ because $3 \times 4 = 12$.

What do you need to multiply $\frac{1}{4}$ by in order to make a denominator of 12? The answer is $\frac{3}{3}$ because $4 \times 3 = 12$.

Multiply $\frac{2}{3}$ by $\frac{4}{4}$ and multiply $\frac{1}{4}$ by $\frac{3}{3}$ to make a LCD of 12:

$\frac{2}{3} = \frac{2\times4}{3\times4} = \boxed{\frac{8}{12}}$ and $\frac{1}{4} = \frac{1\times3}{4\times3} = \boxed{\frac{3}{12}}$

(B) Given $\frac{5}{6}$ and $\frac{3}{10}$, the LCM of 6 and 10 equals 30.

What do you need to multiply $\frac{5}{6}$ by in order to make a denominator of 30? The answer is $\frac{5}{5}$ because $6 \times 5 = 30$.

What do you need to multiply $\frac{3}{10}$ by in order to make a denominator of 30? The answer is $\frac{3}{3}$ because $10 \times 3 = 30$.

Multiply $\frac{5}{6}$ by $\frac{5}{5}$ and multiply $\frac{3}{10}$ by $\frac{3}{3}$ to make a LCD of 30:

$\frac{5}{6} = \frac{5 \times 5}{6 \times 5} = \boxed{\frac{25}{30}}$ and $\frac{3}{10} = \frac{3 \times 3}{10 \times 3} = \boxed{\frac{9}{30}}$

(C) Given $\frac{1}{8}$ and $\frac{7}{12}$, the LCM of 8 and 12 equals 24.

What do you need to multiply $\frac{1}{8}$ by in order to make a denominator of 24? The answer is $\frac{3}{3}$ because $8 \times 3 = 24$.

What do you need to multiply $\frac{7}{12}$ by in order to make a denominator of 24? The answer is $\frac{2}{2}$ because $12 \times 2 = 24$.

Multiply $\frac{1}{8}$ by $\frac{3}{3}$ and multiply $\frac{7}{12}$ by $\frac{2}{2}$ to make a LCD of 24:

$\frac{1}{8} = \frac{1 \times 3}{8 \times 3} = \boxed{\frac{3}{24}}$ and $\frac{7}{12} = \frac{7 \times 2}{12 \times 2} = \boxed{\frac{14}{24}}$

Problems. Express each pair of fractions with their LCD.

① $\frac{1}{2}$ and $\frac{2}{3}$

② $\frac{5}{6}$ and $\frac{4}{9}$

③ $\frac{5}{8}$ and $\frac{7}{10}$

④ $\frac{4}{3}$ and $\frac{5}{4}$

⑤ $\frac{7}{12}$ and $\frac{3}{16}$

⑥ $\frac{6}{7}$ and $\frac{1}{8}$

⑦ $\frac{3}{14}$ and $\frac{2}{21}$

⑧ $\frac{7}{12}$ and $\frac{8}{15}$

⑨ $\frac{5}{7}$ and $\frac{9}{4}$

⑩ $\frac{9}{4}$ and $\frac{7}{8}$

⑪ $\frac{8}{15}$ and $\frac{9}{25}$

⑫ $\frac{7}{24}$ and $\frac{5}{18}$

5.4 Comparing fractions

To compare two fractions, first express each fraction with their LCD (Sec. 5.3). Once the fractions are expressed with the same denominator, whichever fraction has the greater numerator is greater. For example, compare $\frac{5}{6}$ to $\frac{7}{8}$. The LCM of 6 and 8 equals 24. Express $\frac{5}{6}$ and $\frac{7}{8}$ with their LCD to make the equivalent fractions $\frac{5\times4}{6\times4} = \frac{20}{24}$ and $\frac{7\times3}{8\times3} = \frac{21}{24}$. Now that they have the same denominator, we see that $\frac{5}{6} < \frac{7}{8}$ since $\frac{20}{24} < \frac{21}{24}$.

Examples. Use $>$, $<$, or $=$ to compare the given numbers.

(A) Compare $\frac{5}{2}$ and $\frac{7}{3}$. The LCM of 2 and 3 is 6.

$\frac{5}{2} = \frac{5\times3}{2\times3} = \frac{15}{6}$ and $\frac{7}{3} = \frac{7\times2}{3\times2} = \frac{14}{16}$. Since $\frac{15}{6} > \frac{14}{6}$, $\boxed{\frac{5}{2} > \frac{7}{3}}$.

(B) Compare $\frac{1}{6}$ and $\frac{3}{8}$. The LCM of 6 and 8 is 24.

$\frac{1}{6} = \frac{1\times4}{6\times4} = \frac{4}{24}$ and $\frac{3}{8} = \frac{3\times3}{8\times3} = \frac{9}{24}$. Since $\frac{4}{24} < \frac{9}{24}$, $\boxed{\frac{1}{6} < \frac{3}{8}}$.

(C) Compare $\frac{12}{9}$ and $\frac{16}{12}$. The LCM of 9 and 12 is 36.

$\frac{12}{9} = \frac{12\times4}{9\times4} = \frac{48}{36}$ and $\frac{16}{12} = \frac{16\times3}{12\times3} = \frac{48}{36}$. Since $\frac{48}{36} = \frac{48}{36}$, $\boxed{\frac{12}{9} = \frac{16}{12}}$.

(Note that $\frac{12}{9}$ and $\frac{16}{12}$ both reduce to $\frac{4}{3}$.)

Problems. Use >, <, or = to compare the given numbers.

① $\frac{3}{5}$ $\frac{4}{7}$

② $\frac{7}{6}$ $\frac{9}{8}$

③ $\frac{2}{3}$ $\frac{3}{4}$

④ $\frac{3}{10}$ $\frac{4}{15}$

⑤ $\frac{7}{3}$ $\frac{13}{6}$

⑥ $\frac{4}{6}$ $\frac{6}{9}$

Any whole number may be rewritten as a fraction with a denominator of one. For example, $2 = \frac{2}{1}$. The reason is that the fraction line represents division and $\frac{2}{1} = 2 \div 1 = 2$. To compare a whole number to a fraction, first divide the whole number by one. For example, to compare $\frac{11}{3}$ with 4, compare $\frac{11}{3}$ with $\frac{4}{1}$ (since $4 = \frac{4}{1}$).

Example. Use >, <, or = to compare the given numbers.

(D) Compare $\frac{11}{3}$ and 4. First rewrite 4 as $\frac{4}{1}$. Now compare $\frac{11}{3}$ and $\frac{4}{1}$. The LCM of 3 and 1 is 3. Note that $\frac{11}{3}$ already has the LCD. Compare $\frac{11}{3}$ and $\frac{4}{1} = \frac{4\times3}{1\times3} = \frac{12}{3}$. Since $\frac{11}{3} < \frac{12}{3}$, $\boxed{\frac{11}{3} < 4}$.

Problems. Use >, <, or = to compare the given numbers.

⑦ $5 \quad \frac{21}{4}$

⑧ $\frac{19}{6} \quad 3$

⑨ $8 \quad \frac{31}{4}$

An alternative way to compare two fractions is to multiply the numerator of the first fraction with the denominator of the second fraction and multiply the denominator of the first fraction with the numerator of the second fraction.

Example. Use >, <, or = to compare the given numbers.

(E) Compare $\frac{4}{9}$ and $\frac{1}{2}$. Multiply 4 by 2 and multiply 9 by 1.

Compare $4(2) = 8$ and $9(1) = 9$. Since $8 < 9$, $\boxed{\frac{4}{9} < \frac{1}{2}}$.

Problems. Compare using the alternate method.

⑩ $\frac{5}{8}$ $\quad$ $\frac{7}{9}$

⑪ $\frac{8}{9}$ $\quad$ $\frac{9}{10}$

⑫ $\frac{7}{6}$ $\quad$ $\frac{10}{9}$

⑬ $\frac{5}{8}$ $\quad$ $\frac{7}{12}$

5.5 Add and subtract fractions

To add or subtract two fractions, first express each fraction with their LCD (Sec. 5.3). Once the fractions have the same denominator, add or subtract the numerators. If the answer is reducible, reduce your answer (Sec. 5.1).

Examples. Combine the fractions.

(A) $\frac{7}{10} + \frac{4}{15} = \frac{7\times3}{10\times3} + \frac{4\times2}{15\times2} = \frac{21}{30} + \frac{8}{30} = \frac{21+8}{30} = \frac{29}{30}$

(B) $\frac{8}{3} - \frac{6}{5} = \frac{8\times5}{3\times5} - \frac{6\times3}{5\times3} = \frac{40}{15} - \frac{18}{15} = \frac{22}{15}$

Problems. Add the fractions.

① $\frac{2}{3} + \frac{1}{4} =$

② $\frac{5}{6} + \frac{7}{9} =$

③ $\frac{4}{7} + \frac{2}{5} =$

④ $\frac{5}{12}+\frac{7}{20}=$

⑤ $\frac{2}{3}+\frac{1}{9}=$

⑥ $\frac{7}{6}+\frac{9}{8}=$

⑦ $\frac{3}{4}+\frac{1}{12}=$

⑧ $\frac{5}{6}+\frac{1}{10}=$

⑨ $\frac{9}{16}+\frac{7}{24}=$

Problems. Subtract the fractions.

⑩ $\frac{3}{4} - \frac{1}{3} =$

⑪ $\frac{4}{5} - \frac{2}{7} =$

⑫ $\frac{5}{3} - \frac{6}{5} =$

⑬ $\frac{7}{10} - \frac{1}{6} =$

⑭ $\frac{3}{4} - \frac{1}{2} =$

⑮ $\frac{7}{18} - \frac{5}{24} =$

To add or subtract with a whole number and a fraction, divide the whole number by one. For example, $7 = \frac{7}{1}$ since $\frac{7}{1} = 7 \div 1$.

Examples. Add or subtract.

(C) $2 + \frac{1}{4} = \frac{2}{1} + \frac{1}{4} = \frac{2 \times 4}{1 \times 4} + \frac{1}{4} = \frac{8}{4} + \frac{1}{4} = \frac{8+1}{4} = \frac{9}{4}$

(D) $\frac{14}{3} - 4 = \frac{14}{3} - \frac{4}{1} = \frac{14}{3} - \frac{4 \times 3}{1 \times 3} = \frac{14}{3} - \frac{12}{3} = \frac{14-12}{3} = \frac{2}{3}$

Problems. Add or subtract.

⑯ $4 + \frac{2}{3} =$

⑰ $\frac{15}{4} - 2 =$

⑱ $\frac{8}{5} + 7 =$

⑲ $3 - \frac{1}{3} =$

⑳ $5 + \frac{9}{2} =$

5.6 Visualize fraction addition/subtraction

One way to visualize fraction addition and subtraction is to divide a strip into fractions and then divide each fraction based on the LCD. For example, consider $\frac{1}{2}+\frac{1}{3}$. We'll begin by drawing a strip that represents one unit.

Next, we'll divide the unit strip into halves and thirds.

Now we can draw $\frac{1}{2}+\frac{1}{3}$ by joining one half to one third.

What is the LCD? Since the LCM of 2 and 3 is 6, the LCD will be a sixth ($\frac{1}{6}$). Divide $\frac{1}{2}$ into 3 parts and divide $\frac{1}{3}$ into 2 parts (because $2 \times 3 = 6$ and $3 \times 2 = 6$).

This makes a total of five sixths, which shows visually that $\frac{1}{2}+\frac{1}{3}=\frac{5}{6}$.

Examples. (A) Draw $\frac{3}{4} + \frac{1}{6}$ and determine the answer.

Divide one unit into fourths and sixths.

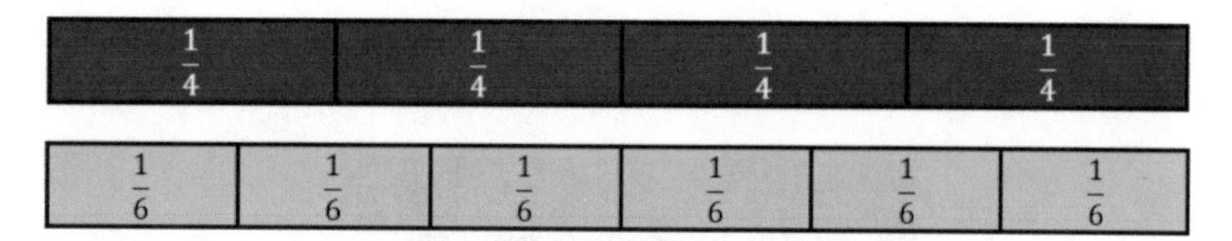

Join three of the fourths to one of the sixths.

Since the LCM of 4 and 6 is 12, the LCD is $\frac{1}{12}$. Divide $\frac{1}{4}$ into 3 parts because $4 \times 3 = 12$ and divide $\frac{1}{6}$ into 2 parts because $6 \times 2 = 12$.

The sum has 11 twelfths. This shows that $\frac{3}{4} + \frac{1}{6} = \frac{11}{12}$.

(B) Draw $\frac{1}{2} - \frac{1}{5}$ and determine the answer.

Divide one unit into halves and fifths.

Draw one of the fifths below one of the halves. The white strip below represents the difference between $\frac{1}{2}$ and $\frac{1}{5}$.

Since the LCM of 2 and 5 is 10, the LCD is $\frac{1}{10}$. Divide $\frac{1}{2}$ into 5 parts because $2 \times 5 = 10$ and divide $\frac{1}{5}$ into 2 parts because $5 \times 2 = 10$.

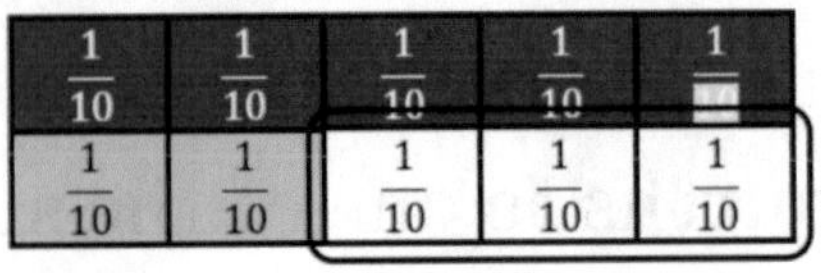

$\frac{1}{10}$	$\frac{1}{10}$	$\frac{1}{10}$	$\frac{1}{10}$	$\frac{1}{10}$
$\frac{1}{10}$	$\frac{1}{10}$	$\frac{1}{10}$	$\frac{1}{10}$	$\frac{1}{10}$

The difference (shown in white) is three tenths. This shows that $\frac{1}{2} - \frac{1}{5} = \frac{3}{10}$.

Problems. Complete each picture and determine the answer.

① $\frac{1}{2} + \frac{2}{5} =$

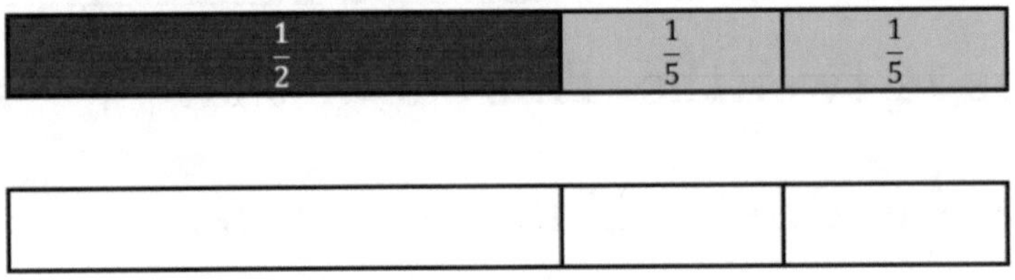

$\frac{1}{2}$	$\frac{1}{5}$	$\frac{1}{5}$

② $\frac{2}{3} - \frac{1}{4} =$

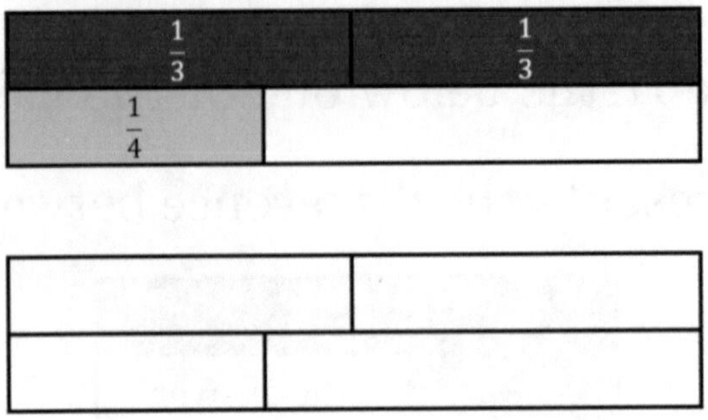

$\frac{1}{3}$	$\frac{1}{3}$
$\frac{1}{4}$	

5.7 Add mixed numbers

To add two mixed numbers, follow these steps:

- Add the fractional parts together.
- If the first step made an improper fraction (meaning that the numerator is larger than the denominator), convert it to a mixed number (Sec. 5.2).
- Add all of the whole numbers together. If the previous step includes a whole number, add that also.
- Combine the whole number and fractional part. Use the fractional part from the second step, if applicable.

Examples. Add the mixed numbers.

(A) $5\frac{1}{4} + 1\frac{2}{3}$ First add the fractional parts: $\frac{1}{4} + \frac{2}{3} = \frac{1\times3}{4\times3} + \frac{2\times4}{3\times4} = \frac{3}{12} + \frac{8}{12} = \frac{11}{12}$ Since $11 < 12$, simply add the whole numbers and the fractional part together: $5\frac{1}{4} + 1\frac{2}{3} = 5 + 1 + \left(\frac{1}{4} + \frac{2}{3}\right) = 5 + 1 + \frac{11}{12} = 6 + \frac{11}{12} = 6\frac{11}{12}$.

(B) $3\frac{1}{2} + 4\frac{2}{3}$ First add the fractional parts: $\frac{1}{2} + \frac{2}{3} = \frac{1\times3}{2\times3} + \frac{2\times2}{3\times2} = \frac{3}{6} + \frac{4}{6} = \frac{7}{6}$ Since $7 > 6$, convert $\frac{7}{6}$ to a mixed number: $\frac{7}{6} = 7 \div 6 = 1\,R1 = 1\frac{1}{6}$. Add all of the whole numbers and the fractional part: $3\frac{1}{2} + 4\frac{2}{3} = 3 + 4 + 1\frac{1}{6} = 3 + 4 + 1 + \frac{1}{6} = 8\frac{1}{6}$.

Problems. Add the mixed numbers.

① $2\frac{3}{5} + 3\frac{1}{4} =$

② $7\frac{1}{2} + 7\frac{1}{3} =$

③ $3\frac{5}{6} + 2\frac{1}{4} =$

④ $2\frac{3}{4} + 1\frac{1}{8} =$

⑤ $4\frac{2}{3} + 7\frac{5}{6} =$

⑥ $9\frac{5}{6} + 8\frac{7}{9} =$

⑦ $4\frac{1}{6} + 2\frac{3}{5} =$

⑧ $5\frac{2}{3} + 6\frac{3}{4} =$

⑨ $4\frac{1}{4} + 7\frac{5}{7} =$

⑩ $1\frac{5}{6} + \frac{3}{8} =$

⑪ $9\frac{2}{3} + 3\frac{7}{9} =$

⑫ $6\frac{4}{7} + 9\frac{3}{8} =$

5.8 Subtract mixed numbers

Note that a mixed number can be "**renamed**" by borrowing one from the whole number. For example, consider $5\frac{3}{4}$. We may borrow one from the whole number and add it to the fractional part: $5\frac{3}{4} = 4 + 1 + \frac{3}{4} = 4 + 1\frac{3}{4} = 4 + \frac{1 \times 4 + 3}{4} = 4 + \frac{7}{4}$. (We converted the mixed number $1\frac{3}{4}$ to $\frac{7}{4}$ using the method from Sec. 5.2.)

To subtract two mixed numbers, follow these steps:

- If the first fractional part is smaller than the second fractional part, rename the first mixed fraction such that its fractional part is an improper fraction. For example, $8\frac{1}{4} = 7 + 1 + \frac{1}{4} = 7 + 1\frac{1}{4} = 7 + \frac{1 \times 4 + 1}{4} = 7\frac{5}{4}$.
- Subtract the whole numbers.
- Subtract the fractional parts. If applicable, use the improper fraction from the first step.
- Combine the answers to the two previous steps to form a mixed number.

Examples. Subtract the mixed numbers.

(A) $6\frac{4}{5} - 3\frac{1}{2}$ Since $\frac{4}{5} > \frac{1}{2}$, simply subtract the whole numbers and subtract the fractional parts:

$6 - 3 = 3$ and $\frac{4}{5} - \frac{1}{2} = \frac{4 \times 2}{5 \times 2} - \frac{1 \times 5}{2 \times 5} = \frac{8}{10} - \frac{5}{10} = \frac{3}{10}$

Combine these answers together: $6\frac{4}{5} - 3\frac{1}{2} = 3\frac{3}{10}$.

(B) $5\frac{1}{3} - 2\frac{3}{4}$ Since $\frac{1}{3} < \frac{3}{4}$, rename the first mixed number:

$5\frac{1}{3} = 4 + 1 + \frac{1}{3} = 4 + 1\frac{1}{3} = 4 + \frac{1 \times 3 + 1}{3} = 4 + \frac{4}{3} = 4\frac{4}{3}$

(We converted $1\frac{1}{3}$ to $\frac{4}{3}$ using the method from Sec. 5.2). The given problem is equivalent to $4\frac{4}{3} - 2\frac{3}{4}$. Subtract the whole numbers and subtract the fractional parts:

$4 - 2 = 2$ and $\frac{4}{3} - \frac{3}{4} = \frac{4 \times 4}{3 \times 4} - \frac{3 \times 3}{4 \times 3} = \frac{16}{12} - \frac{9}{12} = \frac{7}{12}$

Combine these answers together: $5\frac{1}{3} - 2\frac{3}{4} = 2\frac{7}{12}$.

Problems. Subtract the mixed numbers.

① $7\frac{5}{6} - 2\frac{3}{4} =$

② $3\frac{1}{2} - 1\frac{2}{3} =$

③ $1\frac{3}{4} - 1\frac{3}{8} =$

④ $9\frac{2}{5} - 5\frac{4}{7} =$

⑤ $6\frac{1}{4} - 5\frac{1}{3} =$

⑥ $8\frac{4}{9} - 4\frac{1}{6} =$

⑦ $7\frac{2}{7} - \frac{2}{3} =$

⑧ $15\frac{2}{3} - 7\frac{5}{6} =$

An alternative way to subtract mixed numbers is to convert each mixed number into an improper fraction, subtract the improper fractions, and convert the answer into a mixed number. (This idea can also be used to add mixed numbers.)

Example. (C) $5\frac{1}{2} - 2\frac{2}{3} = \frac{5\times2+1}{2} - \frac{2\times3+2}{3} = \frac{11}{2} - \frac{8}{3}$

$= \frac{11\times3}{2\times3} - \frac{8\times2}{3\times2} = \frac{33}{6} - \frac{16}{6} = \frac{33-16}{6} = \frac{17}{6} = 2\frac{5}{6}$

Problems. Subtract using your preferred method.

⑨ $6\frac{2}{5} - 1\frac{3}{4} =$

⑩ $9\frac{3}{8} - 5\frac{7}{12} =$

⑪ $12\frac{8}{15} - 5\frac{4}{9} =$

⑫ $8\frac{1}{6} - 2\frac{5}{8} =$

5.9 Multiply fractions and whole numbers

To multiply a fraction and a whole number together, follow these steps:

- Multiply the numerator and the whole number. Keep the denominator the same.
- If the new numerator and (original) denominator have a common factor, divide the new numerator and the denominator each by the GCF to reduce the fraction.
- If the numerator is evenly divisible by the denominator, the answer is a whole number. For example, $\frac{12}{4} = 3$ since $12 \div 4 = 3$.

Examples. (A) $3 \times \frac{2}{5} = \frac{3\times2}{5} = \frac{6}{5}$ (B) $\frac{3}{4} \times 2 = \frac{3\times2}{4} = \frac{6}{4} = \frac{6\div2}{4\div2} = \frac{3}{2}$

(C) $3 \times \frac{2}{3} = \frac{3\times2}{3} = \frac{6}{3} = 6 \div 3 = 2$

Alternate method: If any numbers in the numerator have a common factor with the denominator, you may cancel the GCF before you multiply. Example B above could have been solved as $\frac{3}{4} \times 2 = \frac{3}{2} \times 1 = \frac{3}{2}$ and Example C above could have been solved as $3 \times \frac{2}{3} = 1 \times \frac{2}{1} = \frac{2}{1} = 2$.

Problems. Multiply the numbers.

① $\frac{4}{5} \times 2 =$

② $8 \times \frac{3}{4} =$

③ $9 \times \frac{2}{3} =$

④ $\frac{7}{6} \times 5 =$

⑤ $\frac{4}{9} \times 6 =$

⑥ $7 \times \frac{7}{2} =$

⑦ $8 \times \frac{5}{2} =$

⑧ $5 \times \frac{7}{3} =$

⑨ $9 \times \frac{5}{12} =$

⑩ $6 \times \frac{5}{6} =$

⑪ $4 \times \frac{7}{8} =$

⑫ $7 \times \frac{1}{7} =$

5.10 Visualize fraction multiplication

If the whole number happens to be evenly divisible by the denominator (and the fraction is a proper fraction), there is a simple way to draw the solution:

- Draw an array of dots where the number of rows equals the denominator and the total number of dots equals the whole number.
- Circle the number of rows equal to the numerator.
- The number of dots circled is the answer.

For example, to draw $\frac{2}{3} \times 15$, draw 15 dots in an array with 3 rows. Circle 2 of the rows. Since 10 dots are circled, the answer is $\frac{2}{3} \times 15 = 10$.

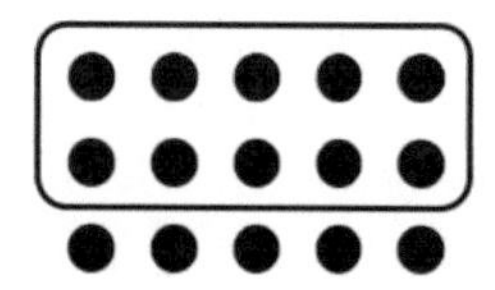

Example. (A) Draw a picture of $\frac{3}{4} \times 16$.

Draw 16 dots in an array with 4 rows. Circle 3 of the rows.

Since 12 dots are circled, the answer is $\frac{3}{4} \times 16 = 12$.

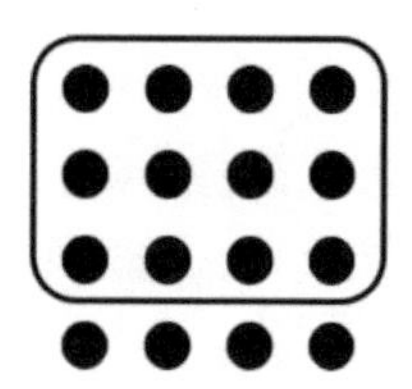

Problems. Draw a picture and determine the answer.

① $\frac{3}{5} \times 20 =$

② $18 \times \frac{1}{3} =$

③ $\frac{5}{6} \times 30 =$

④ $\frac{3}{8} \times 24 =$

Following is another way to draw a fraction multiplied by a whole number (which doesn't require the whole number to be evenly divisible by the denominator):

- Draw horizontal strips of "unit" length to represent the whole number. The number of strips is equal to the whole number.
- If the whole number and denominator share a common factor, divide the denominator by their GCF.
- Divide each strip into equal pieces. The denominator equals the number of pieces. (Use the denominator of the previous step, if applicable.)
- Circle the fraction of the pieces equal to the original fraction (not with the new denominator).
- The answer equals the number of pieces circled over the number of pieces in each strip.

For example, to draw $\frac{5}{6} \times 2$, draw 2 horizontal unit strips. The GCF of 2 and 6 is 2, so the new denominator is $6 \div 2 = 3$. Divide each strip into 3 equal pieces. Each piece represents $\frac{1}{3}$. The original fraction is $\frac{5}{6}$. Circle $\frac{5}{6}$ of the pieces. There are 5 circled pieces and 3 pieces in each strip: The answer is $\frac{5}{3}$.

1			1		
$\frac{1}{3}$	$\frac{1}{3}$	$\frac{1}{3}$	$\frac{1}{3}$	$\frac{1}{3}$	$\frac{1}{3}$

The two horizontal strips at the top represent the whole number 2. We divided each strip into thirds, which made a total of 6 small pieces. We then circled $\frac{5}{6}$ of the pieces. Since each piece equals $\frac{1}{3}$, this shows that $\frac{5}{6} \times 2 = \frac{5}{3}$.

Example. (B) Draw a picture of $\frac{2}{3} \times 4$.

Draw 4 strips. The whole number (4) and denominator (3) don't have any common factors. Divide each strip into 3 pieces. Each piece is $\frac{1}{3}$. Circle $\frac{2}{3}$ of the pieces (2 for every 3). The are 8 circled pieces and 3 pieces in each strip: $\frac{2}{3} \times 4 = \frac{8}{3}$.

1			1			1			1		
$\frac{1}{3}$	$\frac{1}{3}$	$\frac{1}{3}$	$\frac{1}{3}$	$\frac{1}{3}$	$\frac{1}{3}$	$\frac{1}{3}$	$\frac{1}{3}$	$\frac{1}{3}$	$\frac{1}{3}$	$\frac{1}{3}$	$\frac{1}{3}$

(C) Draw a picture of $\frac{3}{4} \times 2$.

Draw 2 strips. The GCF of 2 and 4 is 2, so the new denominator is $4 \div 2 = 2$. Divide each strip into 2 pieces. Each piece is $\frac{1}{2}$. Circle $\frac{3}{4}$ of the pieces (3 for every 4). There are 3 circled pieces and 2 pieces in each strip: $\frac{3}{4} \times 2 = \frac{3}{2}$.

1		1	
$\frac{1}{2}$	$\frac{1}{2}$	$\frac{1}{2}$	$\frac{1}{2}$

Problems. Draw a picture and determine the answer.

⑤ $\frac{3}{5} \times 2 =$

⑥ $4 \times \frac{5}{6} =$

⑦ $\frac{2}{9} \times 6 =$

5.11 Divide fractions and whole numbers

We will consider two types of division problems:

- To divide a whole number by a fraction that has 1 in the numerator (like $2 \div \frac{1}{3}$), multiply the whole number by the denominator. (We'll see why visually in Sec. 5.12.) For example, $2 \div \frac{1}{3} = 2 \times 3 = 6$.
- To divide any fraction by a whole number (like $\frac{1}{3} \div 2$), multiply the whole number by the denominator and write the original numerator over this product. For example, $\frac{1}{3} \div 2 = \frac{1}{3 \times 2} = \frac{1}{6}$ and $\frac{4}{5} \div 3 = \frac{4}{5 \times 3} = \frac{4}{15}$. Also, if the answer can be reduced (Sec. 5.1), reduce it. For example, $\frac{4}{3} \div 8 = \frac{4}{3 \times 8} = \frac{4}{24} = \frac{4 \div 4}{24 \div 4} = \frac{1}{6}$.

Take a moment to compare the two cases above. Although $2 \div \frac{1}{3}$ and $\frac{1}{3} \div 2$ involve the same numbers, the answers of 6 and $\frac{1}{6}$ are much different. You will need to remember how to solve each type of problem.

Examples. (A) $4 \div \frac{1}{3} = 4 \times 3 = 12$ (B) $\frac{2}{3} \div 5 = \frac{2}{3 \times 5} = \frac{2}{15}$

(C) $\frac{6}{7} \div 9 = \frac{6}{7 \times 9} = \frac{6}{63} = \frac{6 \div 3}{63 \div 3} = \frac{2}{21}$

Problems. Divide the numbers.

① $\frac{3}{4} \div 2 =$

② $5 \div \frac{1}{4} =$

③ $\frac{1}{4} \div 5 =$

④ $9 \div \frac{1}{6} =$

⑤ $5 \div \frac{1}{2} =$

⑥ $\frac{6}{5} \div 9 =$

⑦ $6 \div \frac{1}{8} =$

⑧ $\frac{8}{3} \div 4 =$

⑨ $7 \div \frac{1}{7} =$

⑩ $\frac{1}{8} \div 8 =$

⑪ $\frac{9}{8} \div 6 =$

⑫ $4 \div \frac{1}{9} =$

5.12 Visualize fraction division

To draw a whole number divided by a fraction that has 1 in the numerator (like $6 \div \frac{1}{2}$), draw horizontal strips for the whole number and then divide each strip into smaller pieces according to the fraction. For example, to draw $6 \div \frac{1}{2}$, make 6 horizontal strips and then divide each strip in half. The total number of small pieces is the answer. For $6 \div \frac{1}{2}$, dividing 6 strips in half makes a total of 12 small pieces.

1		1		1		1		1		1	
$\frac{1}{2}$	$\frac{1}{2}$	$\frac{1}{2}$	$\frac{1}{2}$	$\frac{1}{2}$	$\frac{1}{2}$	$\frac{1}{2}$	$\frac{1}{2}$	$\frac{1}{2}$	$\frac{1}{2}$	$\frac{1}{2}$	$\frac{1}{2}$

To draw a fraction that has 1 in the numerator divided by a whole number, draw one horizontal strip, divide the strip into pieces according to the denominator, and then divide each piece according to the whole number. The answer is one over the total number of small pieces. For example, to draw $\frac{1}{3} \div 4$, divide a horizontal strip into 3 pieces and then divide each piece into 4 smaller pieces. There are 12 small pieces, so the answer is $\frac{1}{3} \div 4 = \frac{1}{12}$.

$\frac{1}{3}$				$\frac{1}{3}$				$\frac{1}{3}$			
$\frac{1}{12}$	$\frac{1}{12}$	$\frac{1}{12}$	$\frac{1}{12}$	$\frac{1}{12}$	$\frac{1}{12}$	$\frac{1}{12}$	$\frac{1}{12}$	$\frac{1}{12}$	$\frac{1}{12}$	$\frac{1}{12}$	$\frac{1}{12}$

Examples. (A) Draw a picture of $3 \div \frac{1}{2}$.

Draw 3 horizontal strips. Divide each strip into 2 pieces. There were originally 3 pieces. Cutting those pieces in half made a total of 6 small pieces. The answer is $3 \div \frac{1}{2} = 6$.

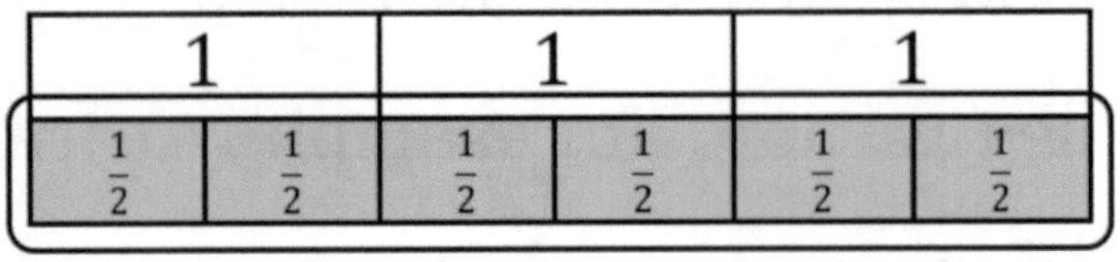

1		1		1	
$\frac{1}{2}$	$\frac{1}{2}$	$\frac{1}{2}$	$\frac{1}{2}$	$\frac{1}{2}$	$\frac{1}{2}$

(B) Draw a picture of $\frac{1}{2} \div 3$.

Draw one horizontal strip. Divide the strip into 2 pieces and then divide each piece into 3 smaller pieces. There are 6 small pieces. When a strip of $\frac{1}{2}$ is divided by 3, the result is a smaller strip equal to $\frac{1}{6}$. The answer is $\frac{1}{2} \div 3 = \frac{1}{6}$.

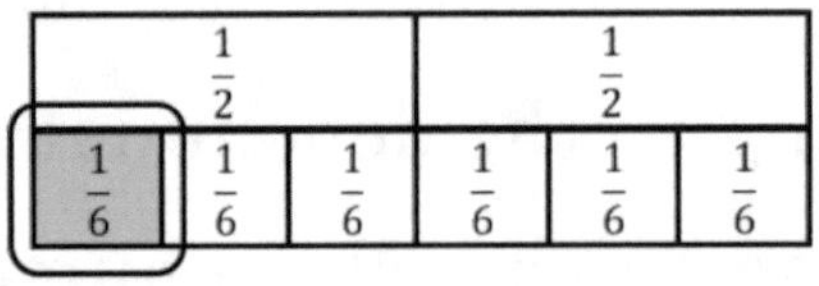

$\frac{1}{2}$			$\frac{1}{2}$		
$\frac{1}{6}$	$\frac{1}{6}$	$\frac{1}{6}$	$\frac{1}{6}$	$\frac{1}{6}$	$\frac{1}{6}$

Problems. Draw a picture and determine the answer.

① $2 \div \frac{1}{3} =$

② $\frac{1}{3} \div 2 =$

5.13 Fractions on the number line

A whole number divided by a fraction with a numerator of one (like $3 \div \frac{1}{5}$) can be used to draw fractional increments on a number line. For example, $3 \div \frac{1}{5}$ corresponds to drawing 3 units on a number line and dividing each unit into fifths, which makes a total of $3 \div \frac{1}{5} = 15$ increments.

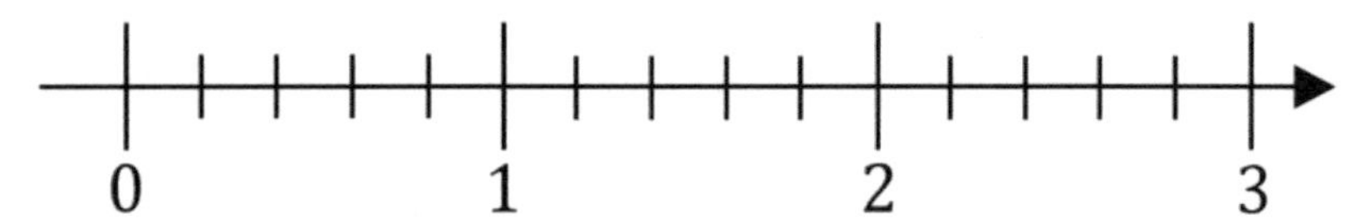

The increments above are $\frac{1}{5}, \frac{2}{5}, \frac{3}{5}, \frac{4}{5}, 1, \frac{6}{5}$, etc. Note that $\frac{5}{5} = 1$, $\frac{10}{5} = 2$, and $\frac{15}{5} = 3$.

Examples. (A) Draw $2 \div \frac{1}{3}$ on a number line. What is $2 \div \frac{1}{3}$?

Draw 2 units. Divide each unit into thirds. Since there are 6 increments all together, $2 \div \frac{1}{3} = 6$.

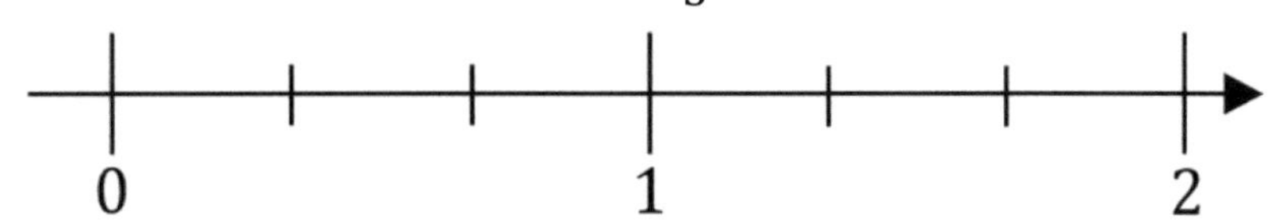

(B) Draw $\frac{1}{3} \div 2$ on a number line. What is $\frac{1}{3} \div 2$?

Draw 1 unit. Divide the unit into thirds. Now divide each increment in half. Since each increment is now $\frac{1}{6}$ of a unit, $\frac{1}{3} \div 2 = \frac{1}{6}$. (See the diagrams at the top of the next page.)

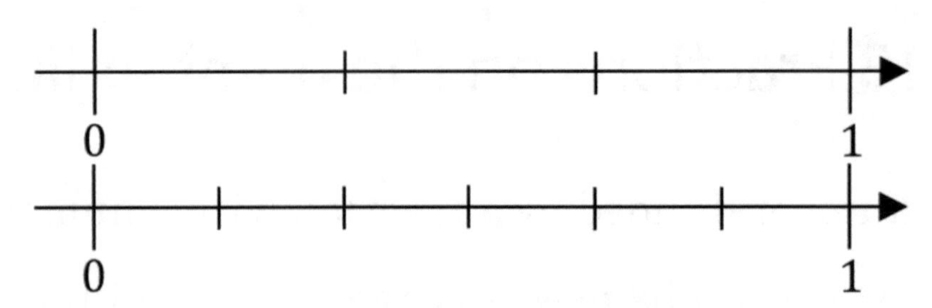

Problems. Draw a number line to answer each question.

① (A) Draw a number line to represent $3 \div \frac{1}{4}$.

(B) Draw and label $\frac{3}{4}$, $\frac{5}{4}$, and $\frac{3}{2}$ on the number line.

(C) Use the number line to determine $3 \div \frac{1}{4}$.

② (A) Draw a number line to represent $\frac{1}{4} \div 3$.

(B) Draw and label $\frac{5}{12}$, $\frac{1}{2}$, and $\frac{5}{6}$ on the number line.

(C) Use the number line to determine $3 \div \frac{1}{4}$.

5.14 Estimating fraction calculations

One way to estimate a calculation involving fractions is to round each fraction to the nearest $\frac{1}{2}$. Here are a few tips:

- A fraction is approximately $\frac{1}{2}$ if the numerator is about half as large as the denominator. For example, $\frac{4}{9}$ is approximately $\frac{1}{2}$ because 4 is almost one-half of 9.
- A fraction is approximately 1 if the numerator is close to the denominator, like $\frac{5}{6}$.
- If the numerator is much smaller than the denominator, like $\frac{1}{8}$, the fraction rounds to zero.
- If the numerator is roughly double the denominator, like $\frac{11}{5}$, the fraction is approximately 2.
- If the denominator plus one-half of the denominator is about equal to the numerator, the fraction is close to $\frac{3}{2}$. For example, $\frac{13}{8}$ is roughly $\frac{3}{2}$ since $8 + \frac{8}{2} = 12 \approx 13$.

This method isn't perfect. It is just intended to provide a rough estimate. The method works better for some cases than for others. (Dividing by a small fraction is a problem.)

Note: This isn't the only method used to estimate fraction calculations. If you want to be more precise, estimate to the nearest $\frac{1}{4}$ or estimate to the nearest $\frac{1}{10}$. In our answer key, we rounded to the nearest $\frac{1}{2}$. Rounding to $\frac{1}{2}$ is simpler. The previous bullet points apply when rounding to $\frac{1}{2}$.

Examples. Estimate the answer.

(A) $\frac{6}{13} + \frac{2}{11} \approx \frac{1}{2} + 0 = \frac{1}{2}$

(B) $\frac{11}{12} - \frac{4}{9} \approx 1 - \frac{1}{2} = \frac{1}{2}$

Problems. Estimate the answer.

① $\frac{6}{11} + \frac{9}{20} \approx$

② $\frac{10}{9} - \frac{17}{35} \approx$

③ $\frac{9}{4} \times \frac{15}{8} \approx$

④ $\frac{19}{12} \div \frac{15}{8} \approx$

⑤ $\frac{20}{9} - \frac{6}{13} \approx$

⑥ $\frac{13}{4} + \frac{9}{5} \approx$

5.15 Word problems with fractions

Reason out which arithmetic operation is involved, or if the solution requires combining operations together. It may be helpful to review Sec.'s 1.9, 2.10, and 4.16.

Tip: In the context of fractions, the word "of" may be used to indicate multiplication. For example, if "$\frac{3}{4}$ of the students have backpacks," to find the number of students who have backpacks, multiply $\frac{3}{4}$ by the total number of students.

Examples. (A) A pitcher originally has $\frac{3}{4}$ gallons of water. If $\frac{1}{6}$ gallons are poured out, how much remains in the pitcher?

Subtract the values to find the difference:

$$\frac{3}{4} - \frac{1}{6} = \frac{3\times3}{4\times3} - \frac{1\times2}{6\times2} = \frac{9}{12} - \frac{2}{12} = \frac{7}{12}$$

The amount remaining is $\frac{7}{12}$ gallons.

(B) There are 16 students in a class. If $\frac{3}{8}$ of the students are wearing blue, how many students are wearing blue?

This is a case where "of" indicates multiplication:

$$\frac{3}{8} \times 16 = \frac{3\times16}{8} = \frac{48}{8} = 48 \div 8 = 6$$

There are 6 students wearing blue.

Problems. Solve each word problem.

① A red string is $\frac{2}{3}$ yards long. A yellow string is $\frac{1}{6}$ yards shorter than the red string. How long is the yellow string?

② If $\frac{3}{4}$ cups of flour, $\frac{1}{3}$ cups of sugar, and $\frac{5}{8}$ cups of water are mixed together, what is the total volume?

③ Charles has 7 bags of sand. Each bag weighs $\frac{3}{4}$ pounds. What is the total weight of the sand?

④ A jug contains $\frac{3}{4}$ gallons of milk. Equal amounts of milk are poured from the jug into 5 bowls. The jug is empty after pouring the milk. How much milk is poured into each bowl?

⑤ Miguel, Lydia, and Eric ordered a pizza with 12 slices. Miguel ate $\frac{1}{4}$ of the pizza, Lydia ate $\frac{1}{6}$ of the pizza, and Eric ate the rest. How many slices did each person eat?

⑥ Liz lives $2\frac{1}{4}$ miles from school. She has already walked $1\frac{5}{6}$ miles from school towards her house. How much farther does she need to walk to reach her home?

⑦ There are 24 students in a class. If $\frac{5}{8}$ of the students are boys, how many of the students are girls?

⑧ A rope is initially 5 feet long. The rope is cut into pieces that are each $\frac{1}{3}$ feet long. How many pieces of rope are there?

⑨ An audio book is $2\frac{1}{2}$ hours long. A boy listened to the book for $\frac{3}{4}$ hours in the morning and $\frac{1}{3}$ hours in the evening. How much time is left in the book?

⑩ A string is initially $\frac{2}{3}$ feet long. The string is cut into 4 equal pieces. How long (in feet) is each piece of string?

⑪ A recipe requires $\frac{5}{8}$ cups of flour to make 4 muffins. How many cups of flour are needed to make one dozen muffins?

⑫ After winning a basketball tournament, the team bought 12 different pies. After the team ate, each pie had $\frac{2}{3}$ slices remaining. How much pie did the team eat all together?

Multiple Choice Questions

① $\frac{28}{49}$ is equivalent to ______.

(A) $\frac{1}{2}$ (B) $\frac{2}{3}$ (C) $\frac{4}{7}$ (D) $\frac{4}{9}$ (E) $\frac{7}{9}$

② Express $4\frac{5}{9}$ as an improper fraction.

(A) $\frac{29}{9}$ (B) $\frac{31}{5}$ (C) $\frac{31}{9}$ (D) $\frac{41}{5}$ (E) $\frac{41}{9}$

③ Express $\frac{29}{6}$ as a mixed number.

(A) $4\frac{1}{2}$ (B) $4\frac{1}{6}$ (C) $4\frac{5}{6}$ (D) $5\frac{1}{2}$ (E) $5\frac{1}{6}$

④ What is the lowest common denominator of $\frac{7}{8}$ and $\frac{5}{12}$?

(A) $\frac{1}{8}$ (B) $\frac{1}{12}$ (C) $\frac{1}{24}$ (D) $\frac{1}{48}$ (E) $\frac{1}{96}$

⑤ Which number is **NOT** equivalent to $\frac{11}{3}$?

(A) $\frac{22}{6}$ (B) $\frac{33}{9}$ (C) $\frac{77}{21}$ (D) $\frac{99}{25}$ (E) $3\frac{2}{3}$

⑥ Which of these numbers is largest?

(A) $\frac{7}{100}$ (B) $\frac{49}{8}$ (C) $\frac{99}{24}$ (D) $4\frac{11}{12}$ (E) $5\frac{3}{4}$

⑦ Order $\frac{17}{6}$, $\frac{9}{4}$, and $2\frac{1}{3}$ from least to greatest.

(A) $\frac{9}{4}, \frac{17}{6}, 2\frac{1}{3}$ (B) $\frac{9}{4}, 2\frac{1}{3}, \frac{17}{6}$ (C) $\frac{17}{6}, \frac{9}{4}, 2\frac{1}{3}$ (D) $2\frac{1}{3}, \frac{9}{4}, \frac{17}{6}$

⑧ What is $\frac{7}{6} + \frac{2}{9}$?

(A) $\frac{1}{2}$ (B) $\frac{3}{5}$ (C) $\frac{10}{9}$ (D) $\frac{25}{6}$ (E) $\frac{25}{18}$

⑨ What is $\frac{11}{12} - \frac{1}{6}$?

(A) $\frac{1}{2}$ (B) $\frac{1}{3}$ (C) $\frac{1}{4}$ (D) $\frac{3}{4}$ (E) $\frac{5}{6}$

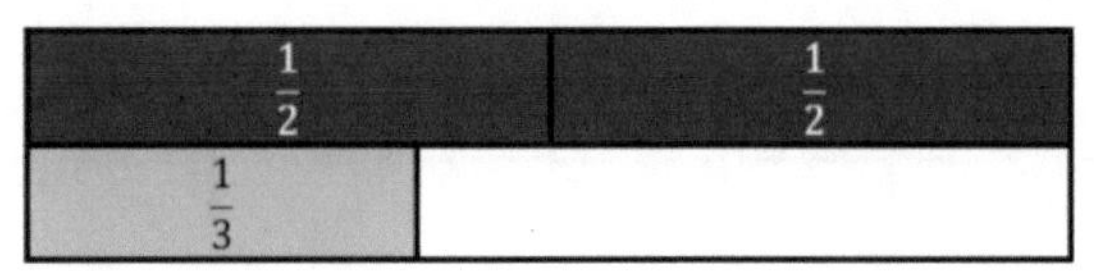

⑩ What is the length of the white strip shown above?

(A) $\frac{1}{6}$ (B) $\frac{2}{3}$ (C) $\frac{3}{4}$ (D) $\frac{5}{6}$ (E) $\frac{5}{8}$

⑪ What is $4\frac{5}{8} + 5\frac{3}{4}$?

(A) $9\frac{2}{3}$ (B) $9\frac{3}{4}$ (C) $9\frac{3}{8}$ (D) $10\frac{2}{3}$ (E) $10\frac{3}{8}$

⑫ What is $7\frac{1}{2} - 4\frac{2}{3}$?

(A) $2\frac{1}{6}$ (B) $2\frac{5}{6}$ (C) $3\frac{1}{6}$ (D) $3\frac{5}{6}$ (E) $4\frac{1}{3}$

⑬ What is $24 \times \frac{5}{6}$?

(A) $\frac{29}{6}$ (B) $\frac{144}{5}$ (C) $\frac{149}{6}$ (D) 20 (E) 120

⑭ Which problem does the diagram above illustrate?

(A) $\frac{1}{4} \times 5$ (B) $\frac{1}{4} \times 15$ (C) $\frac{3}{2} \times 20$ (D) $\frac{3}{4} \times 5$ (E) $\frac{3}{4} \times 20$

⑮ What is $12 \div \frac{1}{4}$?

(A) $\frac{1}{3}$ (B) $\frac{1}{48}$ (C) $\frac{3}{2}$ (D) 3 (E) 48

⑯ What is $\frac{5}{12} \div 6$?

(A) $\frac{5}{2}$ (B) $\frac{5}{72}$ (C) $\frac{1}{2}$ (D) $\frac{2}{5}$ (E) $\frac{72}{5}$

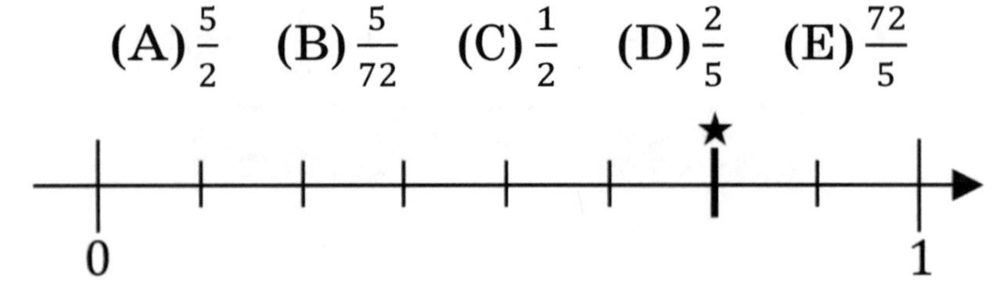

⑰ On the number line above, what is the value of the line that is labeled with a ★?

(A) $\frac{1}{4}$ (B) $\frac{2}{3}$ (C) $\frac{3}{4}$ (D) $\frac{4}{5}$ (E) $\frac{3}{8}$

⑱ If 6 pizzas are divided equally among 8 friends, how much pizza will each friend receive?

(A) $\frac{1}{2}$ (B) $\frac{2}{3}$ (C) $\frac{3}{4}$ (D) $\frac{3}{8}$ (E) $\frac{4}{3}$

⑲ There are 36 fruits and $\frac{4}{9}$ of the fruits are oranges. How many oranges are there?

(A) 12 (B) 16 (C) 18 (D) 20 (E) 81

⑳ A pitcher initially contains $2\frac{1}{4}$ cups of juice. If $\frac{3}{4}$ cups are poured from the pitcher into a glass and $1\frac{1}{2}$ cups are poured from the pitcher into a thermos, how much juice will be left in the pitcher?

(A) 0 (B) $\frac{1}{2}$ cups (C) $\frac{3}{4}$ cups (D) $1\frac{1}{2}$ cups (E) $1\frac{3}{4}$ cups

-6-

DATA ANALYSIS

6.1 Make frequency tables

The **frequency** is a whole number that indicates how many times a particular data value appears in a set of data. For example, consider the table of data below for t-shirts sold. The frequency of M's is 3, the frequency of L's is 5, and the frequency of X's is 2.

M, M, M, L, L, L, L, L, X, X

Frequency is often organized in a table. For example, the table below shows the frequency for the data above. This table indicates that there are 3 M's, 5 L's, and 2 X's.

Number of T-Shirts Sold		
M	L	X
3	5	2

Example. (A) Students were surveyed and their responses are shown below. Make a frequency table for the data.

yes, no, maybe, yes, yes, no, maybe, yes, no

Count the number of occurrences of each data value:

- The frequency of yes is 4.
- The frequency of no is 3.
- The frequency of maybe is 2.

Tip: Add up the frequencies and check that it agrees with the total number of data values: $4 + 3 + 2 = 9$.

The table needs three columns: one for each type of data value (yes, no, and maybe). Organize the frequencies into a table and add a descriptive title at the top.

Student Survey Responses		
yes	**no**	**maybe**
4	3	2

Problems. Make a frequency table for each data set.

① Following are the directions that a car drove:

N, E, W, S, W, N, E, N, W, S, N, W, S, W

② The following tickets were sold at a theater:

child, child, adult, child, child, adult, senior, child, child

6.2 Interpret frequency tables

Frequency tables display the frequencies for categories of data. To determine the total number of data values, add the frequencies together.

Example. (A) A teacher asked students what their favorite color was and organized the results in the following table.

Favorite Color				
red	**blue**	**yellow**	**green**	**pink**
2	7	3	1	4

Which color had the greatest frequency of responses?

Blue had the greatest frequency (7).

Which two colors were the least favorite?

Green (1) and red (2) had the lowest frequencies.

How many different colors were given as answers?

Count the categories: red, blue, yellow, green, and pink.

There are 5 different colors.

How many students were surveyed?

Add the frequencies: $2 + 7 + 3 + 1 + 4 = 17$ students.

Which color is most likely to be a student's favorite?

Blue because it had the greatest frequency (7).

Problems. Use the frequency table to answer the questions.

① The following types of dogs live in a neighborhood.

Dogs in a Neighborhood				
Labrador	**Poodle**	**German Shepard**	**Great Dane**	**Chihuahua**
4	3	5	1	8

(A) Of the types of dogs listed in the table, which type of dog is the most common in the neighborhood?

(B) Of the types of dogs listed in the table, which type of dog is the least common in the neighborhood?

(C) How many dogs live in the neighborhood?

(D) How many different kinds of dogs live in the neighborhood?

(E) If you come across a dog in that neighborhood, which kind of dog will it most likely be?

② A store sold the following items. A book costs $5, a CD costs $10, and a DVD costs $12.

Number of Products Sold		
book	**CD**	**DVD**
15	60	20

(A) How many items did the store sell?

(B) Which type of product was purchased the most?

(C) Which type of product was purchased the least?

(D) How much money did the store receive from DVD sales?

(E) How much money did the store receive overall?

6.3 Draw bar graphs

Frequency data can be visualized by displaying it on a **bar** graph. Follow these steps to draw a bar graph:

- List the categories on the bottom from left to right.
- Choose a suitable scale. A good scale will make it easy to graph all of the numbers. Look at the largest value before you set the scale.
- Number the scale in equal increments vertically.
- Draw a bar for each category based on the frequency.
- Label each axis and add a title for the graph.

Example. (A) Draw a bar graph for the table below.

Weather for the Past Month			
rain	**snow**	**sunny**	**overcast**
10	4	12	4

- Write the categories on the bottom.
- The scale needs to be at least 12 (the highest value). An increment of 2 or 4 would make it easy to draw all of the bars. (We made the increment 4 for this graph.)
- Use the scale to draw bars that match each frequency.
- Label the axes and add a title.

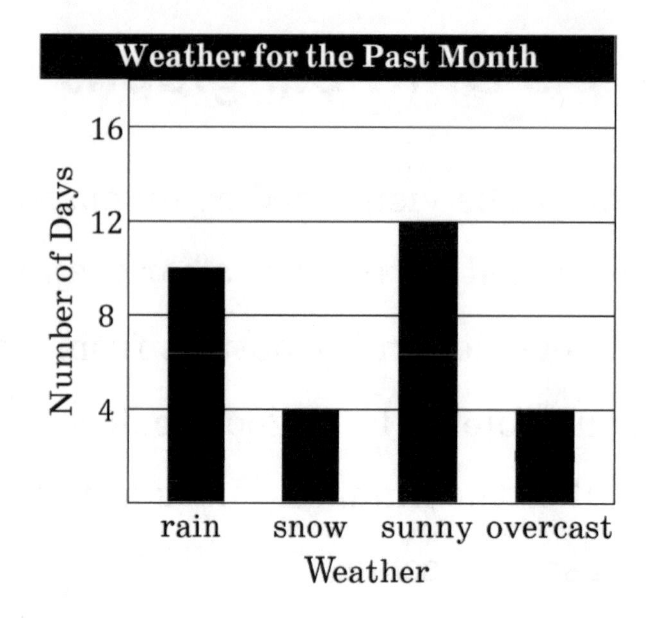

Problems. Draw a bar graph for each frequency table.

① The table below shows the number of games that a team won each month. Draw a bar graph for it.

Games Won in Previous Months				
March	**April**	**May**	**June**	**July**
15	18	24	21	18

② The table below shows which kinds of books have been read at a book club. Draw a bar graph for it.

Book Club Selections			
mystery	**thriller**	**romance**	**horror**
9	6	12	3

③ The table below shows which types of vegetables a restaurant has served. Draw a bar graph for it.

Vegetables Served				
Broccoli	**Carrot**	**Corn**	**Potato**	**Zucchini**
3	1	5	9	2

6.4 Interpret bar graphs

A bar graph is similar to a frequency table except that the data is displayed visually.

Example. (A) Musicians were surveyed to see what their favorite instrument was. The results are displayed below.

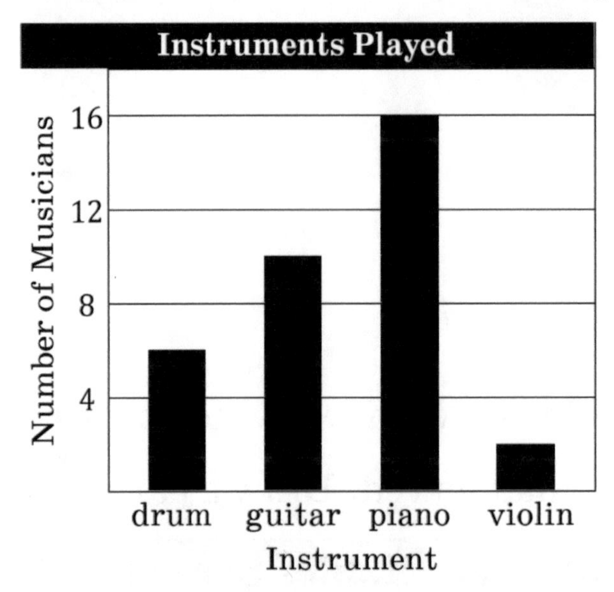

Which instrument was the favorite?

Piano has the greatest frequency (16).

How many more musicians chose the guitar over the drum?

Subtract: $10 - 6 = 4$.

How many musicians were surveyed?

Add the numbers: $6 + 10 + 16 + 2 = 34$ musicians.

Which instrument is least likely to be a favorite?

Violin (based on the sample shown above).

Problems. Use the bar graph to answer the questions.

① The hours a student spent studying is shown below.

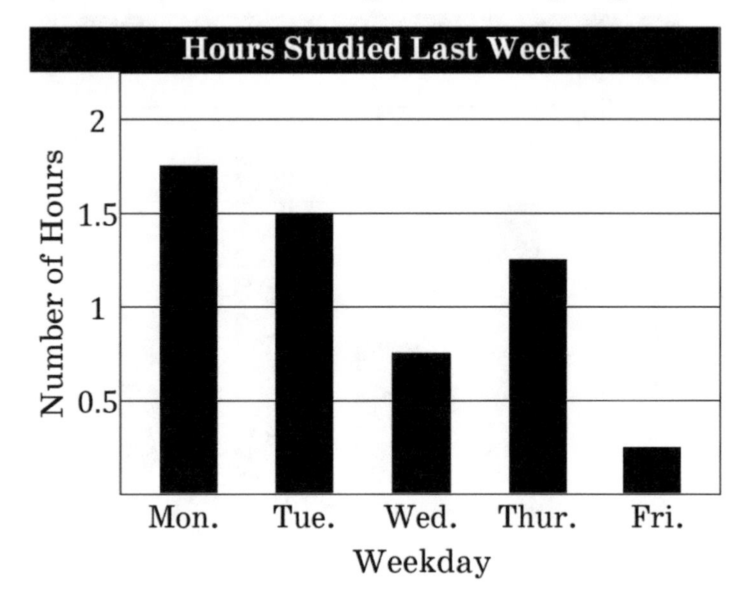

(A) Which day did the student study the most?

(B) How many hours did the student study on Wednesday?

(C) How many hours did the student study all together?

(D) How many more hours did the student study on Tuesday than on Wednesday?

② The cookies sold by a girl scout are shown below. A box of cookies sells for $3.25.

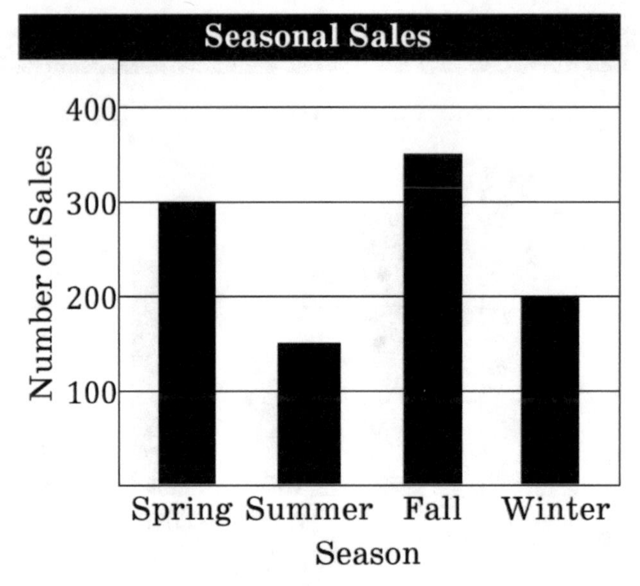

(A) What was the maximum amount of money collected in a single season? Which season was it?

(B) What was the total amount of money collected?

(C) How much more money was collected in the spring than in the summer?

6.5 Draw dot plots

A **dot** plot provides another way to visualize frequency data. Each point is represented by a (●). The dots stack together to form columns. Follow these steps to make a dot plot:

- Draw and label a number line that includes all of the data values. Check the least and greatest values before you label the numbers.
- For each data value, place a dot (●) directly above its value on the number line.

Example. A basketball player recorded the number of free throws that she made in each game this season. Make a dot plot for this data.

8, 6, 9, 8, 7, 5, 8, 7, 6, 8, 9, 7, 5, 8, 6

There are two 5's, three 6's, three 7's, five 8's, and two 9's.

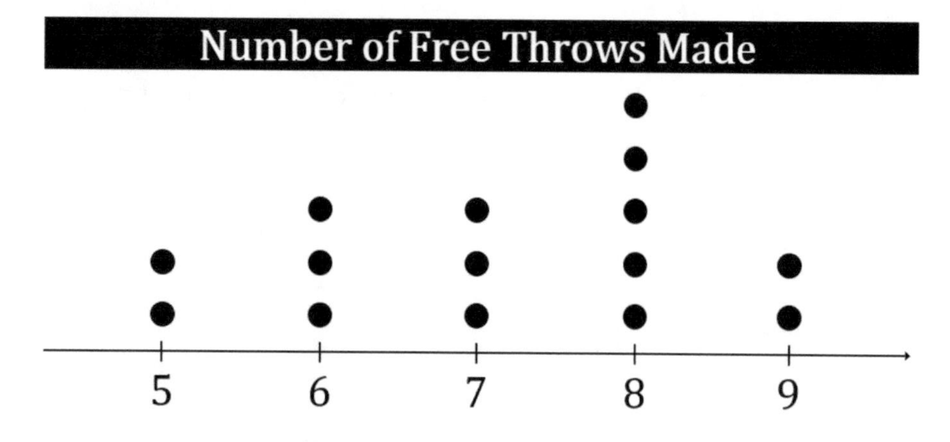

Problems. Draw a dot plot for each data set.

① A boy counts his cards for every turn of a card game:

$$7, 6, 5, 4, 5, 6, 7, 6, 5, 6, 7, 6, 5, 4, 5, 4, 3, 2, 3, 2, 1$$

② A woman keeps track of the number of hours worked:

$$8, 7\frac{1}{2}, 8, 8\frac{1}{2}, 8, 9, 8\frac{1}{2}, 7, 8, 7\frac{1}{2}, 8, 8, 7\frac{1}{2}, 8, 9, 8\frac{1}{2}, 7, 8\frac{1}{2}$$

③ A man records the weights of bowling balls in pounds:

$$11, 12.5, 12, 11.5, 12, 12, 12.5, 11.5, 12, 12.5, 12.5, 11, 12$$

6.6 Range

The **range** of a data set equals the difference between the maximum value and the minimum value. The range provides a measure of the full spread of the data. A data set with a small range is spread narrowly, whereas a data set with a large range is spread widely.

Examples. Find the range.

(A) For 7, 5, 11, and 10, the range is $11 - 5 = 6$.

(B) For 45, 31, 21, 39, and 28, the range is $45 - 21 = 24$.

Problems. Find the range for each data set.

① 8, 7, 5, 3, 9, 7, 4, 6, 8, 5, 7, 4, 8, 6

② 9, 12, 6, 15, 24, 9, 18, 15, 12, 21, 9, 18, 15, 21, 18, 9, 12

③ 5, 5, 5, 5, 5

④ 3.6, 5.2, 4.8, 1.7, 9.3

⑤ $4\frac{3}{4}, 3\frac{1}{2}, 5\frac{1}{4}, 2\frac{3}{4}, 6\frac{2}{3}, 4\frac{1}{4}, 7\frac{1}{3}, 5\frac{3}{4}, 4\frac{1}{2}$

⑥ 0.0375, 0.085, 0.0625, 0.09, 0.045, 0.0575

⑦ $\frac{3}{8}, \frac{9}{4}, \frac{2}{7}, \frac{6}{5}$

6.7 Interpret dot plots

A dot plot has numerical values on the bottom and displays data as individual dots.

Example. (A) The ages of bus passengers are shown below.

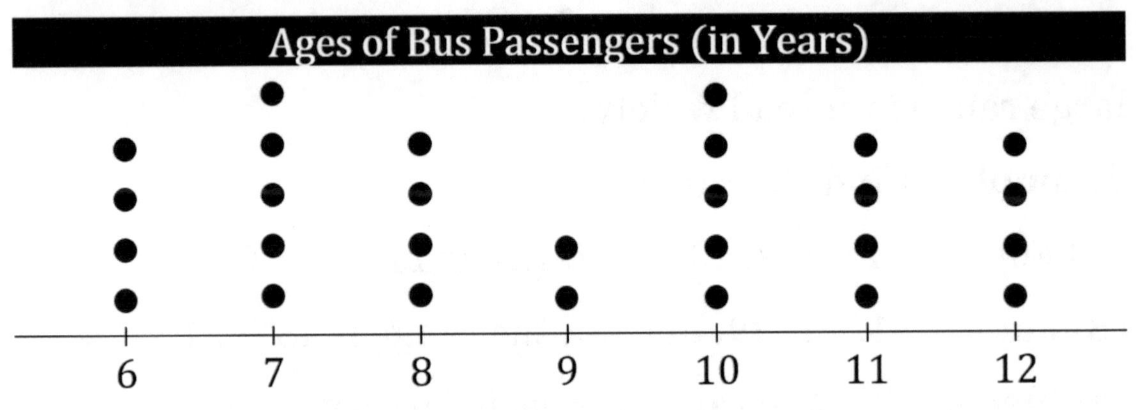

What is the range of the passengers' ages?

The smallest age is 6 and the largest age is 12. Subtract these to find the range: $12 - 6 = 6$.

How many passengers rode on the bus?

Count the dots: $4 + 5 + 4 + 2 + 5 + 4 + 4 = 28$.

What is the sum of all of the passengers' ages?

Add up their ages:

$$6 \times 4 + 7 \times 5 + 8 \times 4 + 9 \times 2 + 10 \times 5 + 11 \times 4 + 12 \times 4$$
$$= 24 + 35 + 32 + 18 + 50 + 44 + 48 = 251$$

Which two ages are most common on the bus?

7 years and 10 years.

Problems. Use the dot plot to answer the questions.

① The plot below shows water consumed.

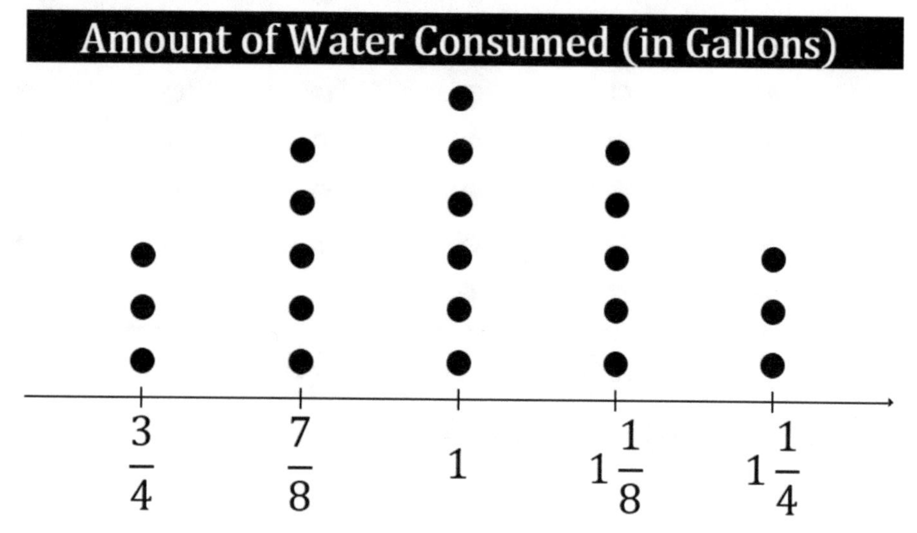

(A) What is the range of water consumed?

(B) How many data points are there in total?

(C) What is the total amount of water consumed?

② The plot below shows the cost of items in dollars.

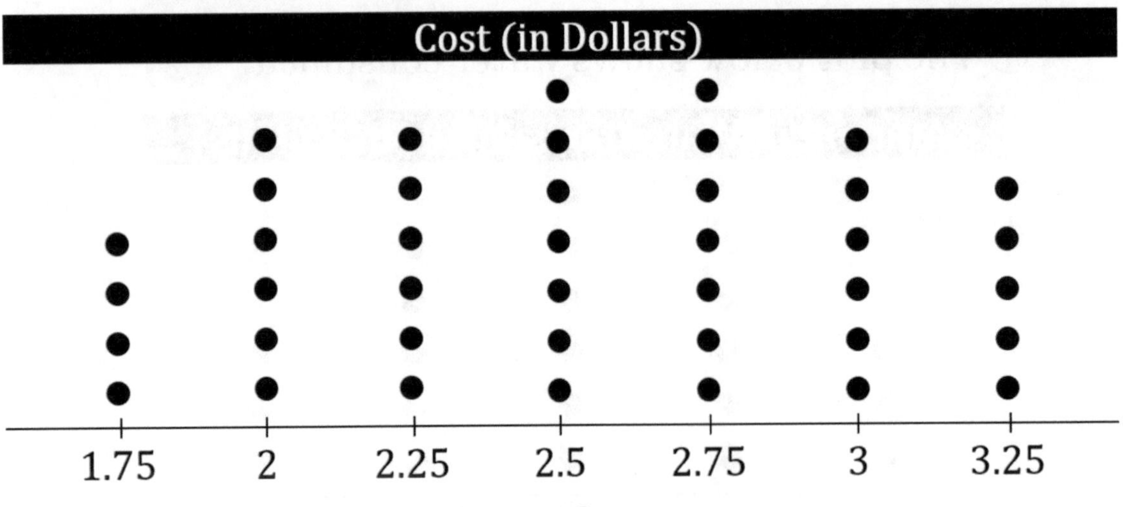

(A) What is the range of costs?

(B) How many more items cost $2.75 than $1.75?

(C) What is the total number of items?

(D) What is the total cost of all the items?

6.8 Draw stem-and-leaf plots

A **stem-and-leaf** plot provides a concise way to display data when all of the values have multiple digits. For example, a stem-and-leaf plot with two-digit numbers treats the tens digit as the **stem** and the units digit as the **leaf**. This allows a set of two-digit numbers to be grouped together. For example, the numbers 32, 33, 35, 37, and 38 can be grouped together by separating the tens digit (3) from the units digit. The row of numbers below represents 32, 33, 35, 37, and 38.

3	2 3 5 7 8

A stem-and-leaf plot organizes data by stems and leaves. To make a stem-and-leaf plot, follow these steps:

- Make stems and leaves. For two-digit numbers, use the tens digit as the stem and the units digit as the leaf.
- List each stem in the left column. Place each stem in its own row (so the stems are stacked up vertically).
- For each stem, write the corresponding leaves in order from least to greatest in the right column.
- Note that a value may be repeated. (See the example.)
- Use one stem and leaf for a key, like 3|2 represents 32.

Example. Make a stem-and-leaf plot for the football scores shown below.

$$20, 35, 12, 30, 21, 34, 27, 17, 24, 24, 33, 14, 21, 38, 17, 35, 28, 14$$

First arrange the data in order from least to greatest.

$$12, 14, 14, 17, 17$$

$$20, 21, 21, 24, 24, 27, 28$$

$$30, 33, 34, 35, 35, 38$$

The stems are the tens digits: 1, 2, and 3. Every number consists of a stem and a leaf. For example, in 12 the stem is 1 and the leaf is 2. Arrange the stems in the left column. For each stem, write the corresponding leaves (these are the units digits) in the right column. Note that the values in the stem-and-leaf plot below are identical to the values listed above.

Football Scores	
Stems	Leaves
1	2 4 4 7 7
2	0 1 1 4 4 7 8
3	0 3 4 5 5 8
1|4 represents 14.	

Put the key at the bottom: "1|4 represents 14" is an example to show that tens and units digits are stems and leaves.

Problems. Draw a stem-and-leaf plot for each data set.

① Weekly high temperatures in Fahrenheit include:

88, 73, 92, 80, 67, 74, 68, 60, 54, 59, 70, 74, 81, 86

② The ages of people at a book club are:

32, 25, 36, 65, 32, 28, 41, 38, 42, 30, 36, 27, 32, 39

③ A basketball team earned the following scores:

95, 103, 92, 100, 96, 114, 90, 106, 89, 110, 101

6.9 Interpret stem-and-leaf plots

A stem-and-leaf plot organizes data by stems and leaves, which are usually tens digits and units digits.

Example. (A) The attendance for a class is shown below.

Daily Attendance Totals	
Stems	**Leaves**
2	3 3 6 8
3	1 3 5 7 7 7 8 9
4	0 2 3 3
	2|3 represents 23.

What is the range of attendance totals?

The smallest is 23 and the largest is 43. Subtract these to find the range: $43 - 23 = 20$.

Which attendance total occurred most often?

37 occurred three times.

On how many days was attendance taken?

Count the data values. There are 4 in the 20's, 8 in the 30's, and 4 in the 40's: $4 + 8 + 4 = 16$.

On how many days was the attendance total greater than 40?

Three days: 42, 43, and 43.

Problems. Use the plot to answer the questions.

① The plot below shows the ages of people at a camp.

Ages at a Camp (in Years)	
Stems	**Leaves**
0	7 7 7 8 8 8 8 9 9 9 9 9
1	0 1 1 1 2 2 2 2 2 2 3 3 3 4 4
2	1 1 2 2
	2|1 represents 21.

(A) What are the ages of the youngest and oldest people?

(B) Which age is most common?

(C) How many people at the camp are older than twelve?

(D) How many people were at the camp?

② The plot below shows money collected from sales.

Money Collected (in Dollars)	
Stems	**Leaves**
3	6 8
4	0 0 4 5 5 5 5 8 8
5	0 0 0 5 5 5 6 8
6	5 5 8 8
7	0 2 2 5
	3|6 represents 36.

(A) What is the range of values of money collected?

(B) Which amount was most common?

(C) On how many occasions was money collected?

(D) Estimate the total amount of money collected.

6.10 Make scatter plots

A **scatter** plot graphs ordered pairs to show the relationship between two quantities. An **ordered pair** is a pair of numbers in parentheses like $(1, 3)$ or $(2, 6)$. An ordered pair locates a point on the **coordinate plane**. The coordinate plane has two axes: the x-axis is horizontal and the y-axis is vertical. The x- and y-axes intersect at the point $(0, 0)$, called the **origin**. The first coordinate (x) of the ordered pair (x, y) indicates how far the point is to the right of the origin, and the second coordinate (y) indicates how far the point is above the origin. For example, the point $(6, 2)$ is 6 units to the right and 2 units up, as shown below.

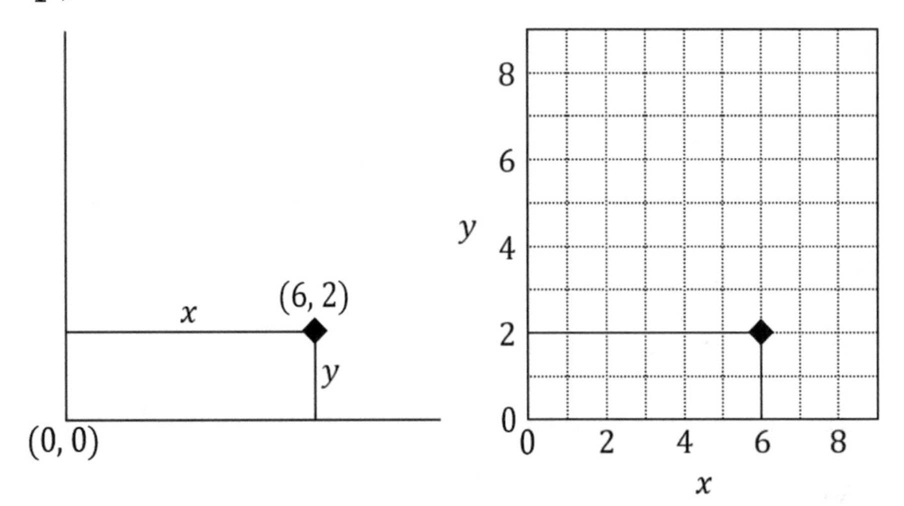

The **independent** variable is plotted on the x-axis (which is horizontal) and the **dependent** variable is plotted on the y-axis (which is vertical). The independent variable is either something that changes on its own (like the temperature outside) or that can be manipulated or controlled (like the number of games played). The dependent variable depends on what happens to the independent variable. For example, if a person earns $10 per hour, the number of hours that the person works is the independent variable and the amount of money earned is the dependent variable (because we need to know how many hours the person works in order to figure out how much money the person earns).

Follow these steps to make a scatter plot:

- Identify the independent and dependent variables.
- Choose a suitable scale for each axis based on the least and greatest values for each variable.
- Graph each ordered pair with the independent variable on the horizontal axis and the dependent variable on the vertical axis.
- Label each axis and add a title.

Example. (A) A girl uses a spring to launch a small object up into the air. She measures the distance that the spring is compressed and the height that the object rises into the air, as tabulated below. Draw a scatter plot for this data.

Spring Launch Data				
Compression (cm)	2	4	6	8
Height (m)	0.5	1	3	7

The height that the object rises upward depends on how far the spring is compressed. The independent variable is the compression of the spring and the dependent variable is the height that the object rises. Put the compression on the horizontal (x) axis and the height on the vertical (y) axis. For example, when the compression is 2 cm, the height is 0.5 m. To plot this point, go 2 units right and 0.5 m up.

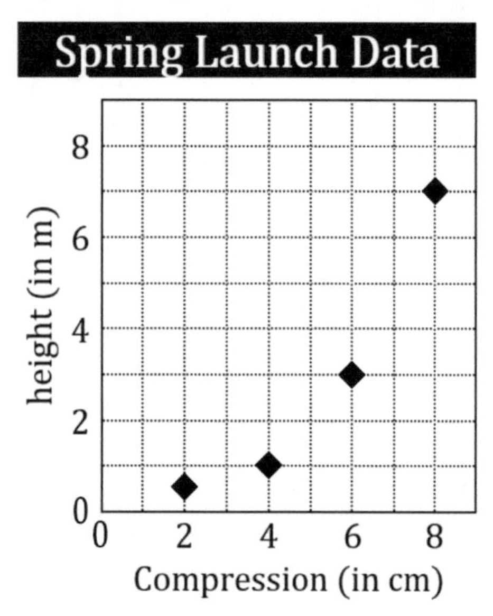

Problems. Draw a scatter plot for each set of data.

① Draw a scatter plot for the attendance data below.

Weekly Attendance					
Time (weeks)	0	4	8	12	16
Class Attendance	40	35	24	18	22

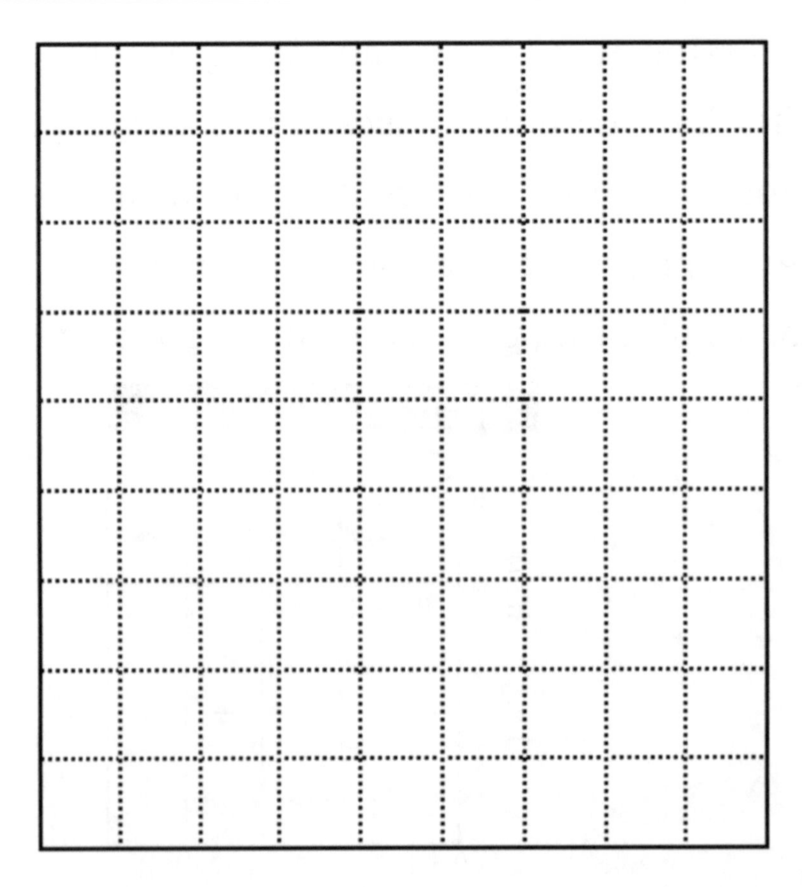

② Draw a scatter plot for the used car values below.

Used Car Value					
Value ($1000)	12	8	5	3	2
Age (years)	0	2	4	6	8

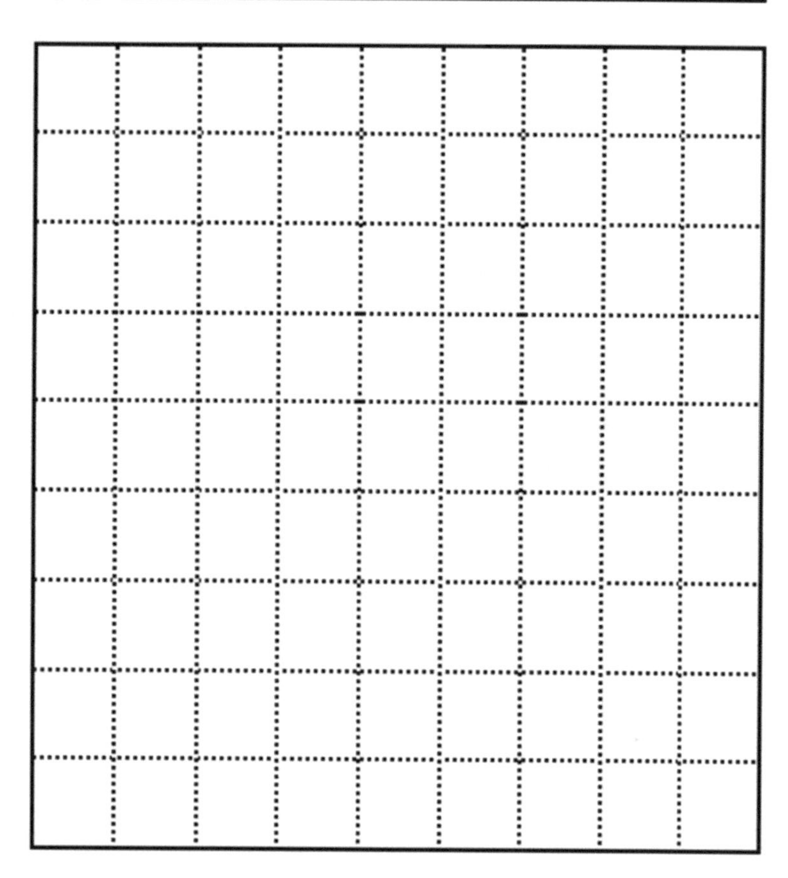

③ Draw a scatter plot for the umbrella sales below.

Umbrella Sales					
Rainfall (in.)	0	0.5	1	1.5	2
Number of Sales	0	2	8	12	19

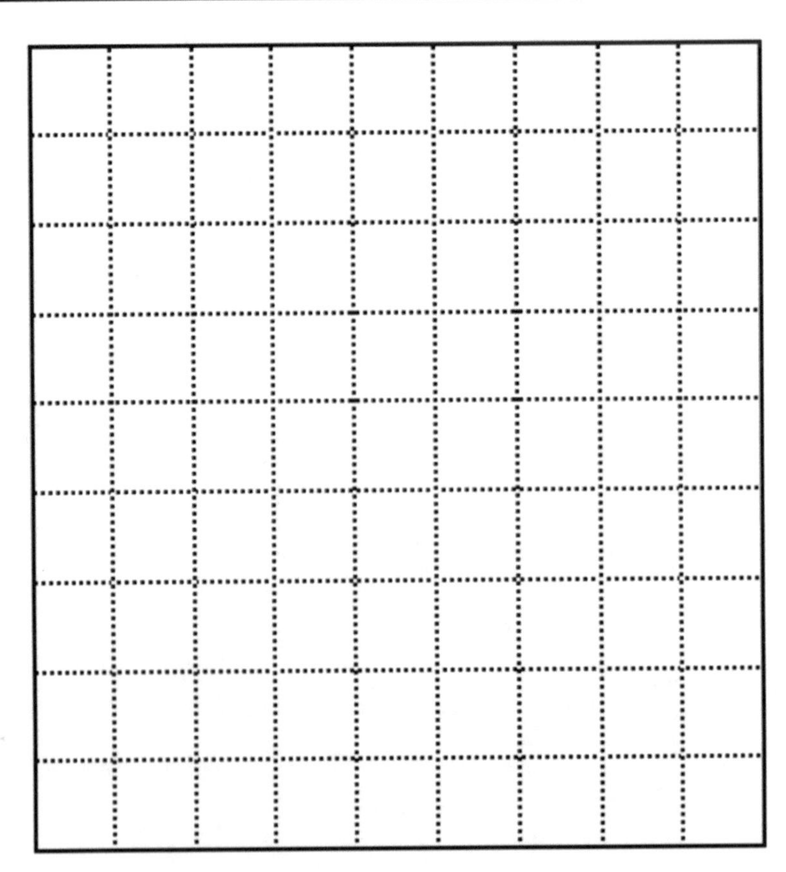

6.11 Interpret scatter plots

A scatter plot shows the relationship between the quantities on the horizontal and vertical axes.

- If y increases as x increases, the quantities are **directly** proportional.
- If an increase in x is accompanied by a decrease in y, the quantities are **inversely** proportional.
- If the graph appears fairly random and scattered, x and y don't have a relationship.

Example. **(A)** Data for a circuit is shown below.

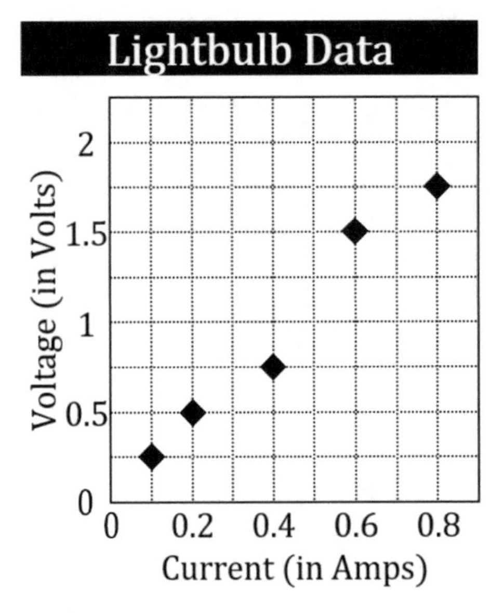

What was the voltage when the current was 0.6 amps?

Find 0.6 amps on the current, find the data point above it, and read off the voltage: $\boxed{1.5}$ volts.

According to the graph, what type of relationship do current and voltage have?

They appear to be directly proportional. As the current increases, so does the voltage. For example, when the current increases from 0.2 amps to 0.8 amps, the voltage increases from 0.5 volts to 1.75 volts.

How many measurements were made?

Count the data points. There are 5 data points on the graph.

Which variable is the dependent variable?

Voltage is on the vertical axis.

Which variable is the independent variable?

Current is on the horizontal axis.

If the current is 0.5 amps, about how much voltage would you expect to measure?

There isn't a data point for 0.5 amps, but there are data points for 0.4 amps and 0.6 amps. We can use these to make a prediction. The corresponding voltages are 0.75 volts and 1.5 volts. The answer must lie somewhere in between these values. A good *estimate* is the average value in between of approximately 1.1 volts.

Problems. Use the scatter plot to answer the questions.

① The plot below shows earnings for a part-time job.

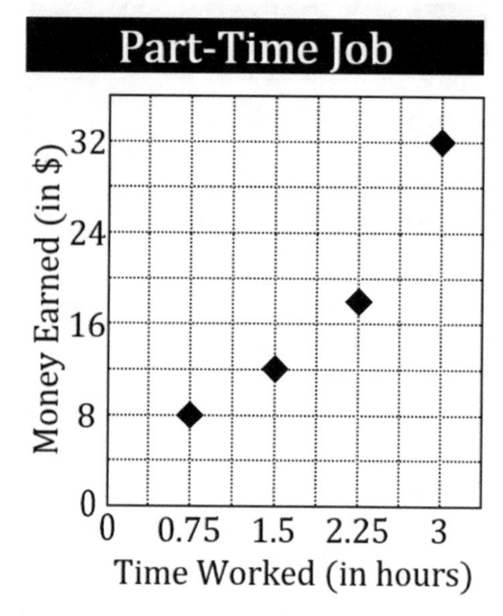

(A) Which type of relationship do the data follow?

(B) How much money was earned for 1.5 hours of work?

(C) How much time was spent working to earn $18?

(D) Predict the earnings for two hours of work.

② The plot below shows a person's golf scores on a nine-hole golf course.

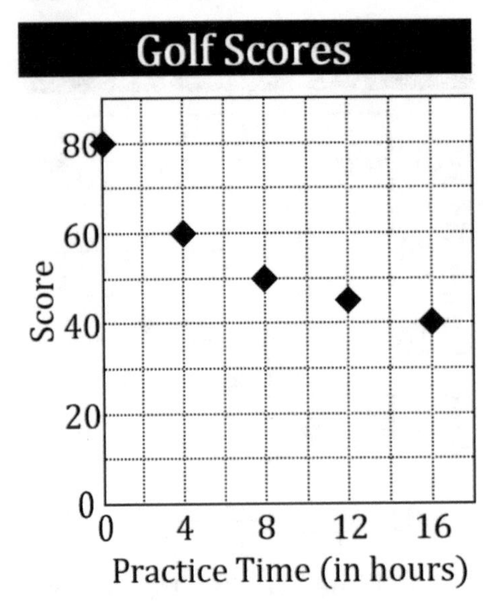

(A) Which type of relationship do the data follow?

(B) How many scores were less than 50?

(C) What was the score for 12 hours of practice?

(D) Predict the score for two hours of practice.

Multiple Choice Questions

① Which type of graph is best suited for the data below?

68, 70, 64, 75, 83, 77, 68, 75, 81, 72

(A) bar graph (B) dot plot (C) scatter plot (D) stem-and-leaf

② Which type of graph is best suited for the data below?

5, 2, 3, 5, 2, 7, 6, 4, 5, 2, 3, 2, 5, 6, 5, 4

(A) bar graph (B) dot plot (C) scatter plot (D) stem-and-leaf

③ Which type of graph is best suited for the data below?

Sizes of Drinks Purchased			
small	medium	large	XL
27	35	42	19

(A) bar graph (B) dot plot (C) scatter plot (D) stem-and-leaf

④ Which type of graph is best suited for the data below?

Scale Measurements					
Weight (N)	4.8 N	9.8 N	14.9 N	19.3 N	24.5 N
Mass (kg)	0.5	1	1.5	2	2.5

(A) bar graph (B) dot plot (C) scatter plot (D) stem-and-leaf

⑤ Every year, a mother measures the height of her daughter. What is the independent variable?

(A) distance (B) height (C) mass (D) size (E) time

Hours of Exercise				
January	February	March	April	May
24	18	21	15	12

⑥ In the table above, the frequency is highest in ______.

(A) Jan. (B) Feb. (C) Mar. (D) Apr. (E) May

⑦ In the table above, how many total hours of exercise occurred in March and April?

(A) 6 (B) 9 (C) 27 (D) 33 (E) 36

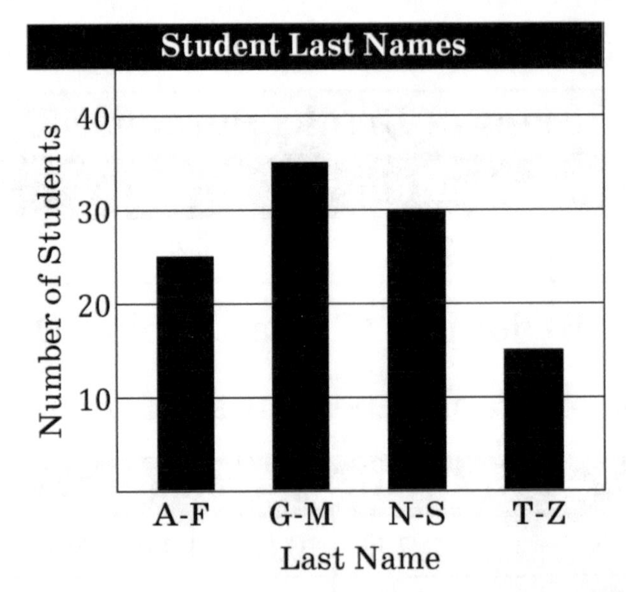

⑧ On the graph above, how many last names are in G-M?

(A) 20 (B) 25 (C) 30 (D) 35 (E) 40

⑨ On the graph above, how many more last names are in A-F than are in T-Z?

(A) 5 (B) 10 (C) 15 (D) 20 (E) 40

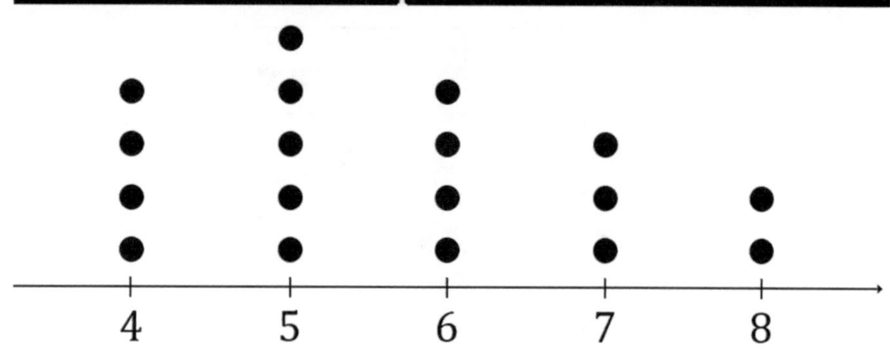

⑩ On the plot above, what is the range?

(A) 4 (B) 5 (C) 6 (D) 7 (E) 8

⑪ On the plot above, what is the frequency of 6's?

(A) 2 (B) 3 (C) 4 (D) 5 (E) 6

⑫ On the plot above, how many runs were scored in total?

(A) 8 (B) 25 (C) 30 (D) 92 (E) 102

Number of Pages Read	
Stems	Leaves
1	4 7 8
2	1 3 4 4 4 5 5 7
3	0 1 1 6
	1|4 represents 14.

⑬ On the graph above, which page total occurred the most?

(A) 14 (B) 24 (C) 25 (D) 31 (E) 36

⑭ On the graph above, how often were over 30 pages read?

(A) 1 time (B) 2 times (C) 3 times (D) 4 times (E) 6 times

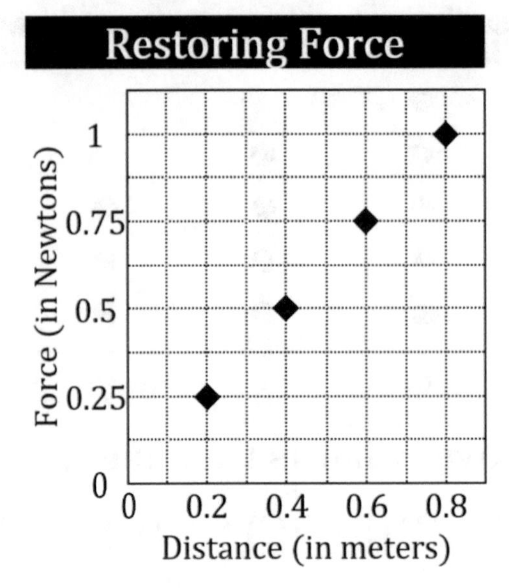

⑮ On the graph above, what is the force when the distance is 0.6 m?

(A) 0 N (B) 0.25 N (C) 0.5 N (D) 0.75 N (E) 1 N

⑯ On the graph above, the relationship is ___________.

(A) directly proportional

(B) inversely proportional

(C) totally random

⑰ On the graph above, predict the force when the distance is 0.5 m.

(A) 0.25 N (B) 0.5 N (C) 0.625 N (D) 0.75 N (E) 1 N

⑱ The graph above is a ___________.

(A) bar graph (B) dot plot (C) scatter plot (D) stem-and-leaf

-7-

NUMBER PATTERNS

7.1 Additive patterns

An additive pattern is based on addition or subtraction. The pattern is based on the difference between the numbers. For example, any two consecutive numbers in the list below have a difference of 6.

$$6, 12, 18, 24, 30, 36, 42, \dots$$

You should recognize the numbers above as multiples of 6. For example, $6 \times 2 = 12$ and $6 \times 3 = 18$. However, numbers don't need to be multiples to form an additive pattern. For example, the pattern below involves a difference of 4, and these numbers don't form a list of multiples.

$$7, 11, 15, 19, 23, 27, 31, \dots$$

Examples. Fill in the missing numbers.

(A) 8, 16, 24, 32, 40, 48, ____, ____ Add 8 to each number to get the next number. For example, $32 + 8 = 40$ and $40 + 8 = 48$. The next two numbers are $48 + 8 = 56$ and $56 + 8 = 64$.

(B) 51, 47, 43, ____, 35, 31, 27, ____ Subtract 4 each time. For example, $51 - 4 = 47$ and $47 - 4 = 43$. The missing numbers are $43 - 4 = 39$ and $27 - 4 = 23$.

Problems. Fill in the missing numbers.

① 9, 18, 27, 36, 45, 54, _____, _____

② 75, 67, _____, 51, 43, 35, 27, _____

③ 8, 20, _____, 44, 56, 68, _____, 92

④ 159, _____, 121, 102, 83, _____, 45, 26

⑤ 1, 2, 4, _____, 11, 16, _____, 29

⑥ 8, 15, 24, 31, 40, _____, 56, _____

⑦ 0.125, 0.25, _____, 0.5, _____, 0.75, 0.875, 1

⑧ $\frac{5}{16}, \frac{3}{8}, \frac{7}{16},$ _____, $\frac{9}{16}, \frac{5}{8}, \frac{11}{16},$ _____

⑨ $\frac{3}{4}, 2, 3\frac{1}{4}, 4\frac{1}{2},$ _____, _____, $8\frac{1}{4}, 9\frac{1}{2}$

⑩ 5, 4, 5, 5, 5, 6, 5, 7, _____, _____

7.2 Multiplicative patterns

A multiplicative pattern is based on multiplication or division. The pattern is based on multiplying or dividing two numbers. For example, any two consecutive numbers in the list below are different by a factor of 2. For example, $16 \times 2 = 32$ and $32 \times 2 = 64$.

$$1, 2, 4, 8, 16, 32, 64, \ldots$$

Examples. Fill in the missing numbers.

(A) $2, 10, 50, 250, \_\_\_\_, \_\_\_\_$ Multiply each number by 5 to get the next number. For example, $2 \times 5 = 10$ and $10 \times 5 = 50$. The next two numbers are:

- $250 \times 5 = 1250$
- $1250 \times 5 = 6250$

(B) 1,000,000, 100,000, ________, 1000, 100, ________ Divide by 10 each time. For example, $1{,}000{,}000 \div 10 = 100{,}000$ and $1000 \div 10 = 100$. The missing numbers are:

- $100{,}000 \div 10 = 10{,}000$
- $100 \div 10 = 10$

Problems. Fill in the missing numbers.

① 2, 6, 18, 54, _____, _____

② 4802, _____, 98, 14, _____

③ 0.02, 0.08, 0.32, ______, 5.12, ______

④ 1000, _____, 10, 1, 0.1, _____

⑤ $\frac{1}{8}$, _____, $\frac{1}{2}$, 1, 2, _____

⑥ 12, 6, 3, _____, _____, $\frac{3}{8}$

⑦ 4, 12, 48, 240, ________, ________

⑧ $\frac{2}{3125}$, $\frac{2}{625}$, _____, $\frac{2}{25}$, $\frac{2}{5}$, _____

⑨ $\frac{1}{720}$, $\frac{1}{120}$, $\frac{1}{24}$, _____, $\frac{1}{2}$, _____

⑩ $\frac{729}{64}$, $\frac{243}{32}$, _____, $\frac{27}{8}$, $\frac{9}{4}$, _____

7.3 Prime and composite numbers

A **prime number** is a whole number greater than 1 which is evenly divisible only by two whole numbers: 1 and itself. A **composite number** is a whole number greater than 1 which isn't a prime number. It may help to review the divisibility tests from Sec. 2.14. For example, 46 is evenly divisible by 2 because 46 is even, and 594 is evenly divisible by 9 because $5 + 9 + 4 = 18$ is a multiple of 9. To determine if a number is prime, you need to check all of the prime numbers up to its square root. For example, to tell whether or not 143 is prime, you need to check whether it is evenly divisible by 2, 3, 5, 7, and 11. You don't need to check 4, 6, 8, 10, or 12 because if it isn't evenly divisible by 2, it won't be evenly divisible by any other even numbers. You don't need to check 9 because if it isn't evenly divisible by 3, it won't be evenly divisible by 9. Since $12 \times 12 = 144$, you don't need to check numbers higher than 11. Why not? If a number greater than its square root evenly divides into it, the other factor will be less than its square root. For example, $161 \div 23 = 7$ would have been discovered when testing $161 \div 7 = 23$.

Example. Fill in the missing numbers.

(A) 2, 3, 5, 7, 11, 13, ____, ____ These are prime numbers. The next two numbers are 17 and 19. (Note that 9 and 15 were "skipped" since they are composite: $3 \times 3 = 9$ and $3 \times 5 = 15$.)

Problems. Fill in the missing numbers.

① 23, 29, 31, 37, 41, 43, _____, _____

② 2, 5, 11, _____, 23, 31, 41, _____

③ 97, 89, 83, _____, 73, 71, _____, 61

④ 9, 15, 21, 25, 27, 33, _____, _____

⑤ 1, 2, 4, 6, 10, 12, _____, _____

⑥ 0.8, 1.2, 2, 2.8, _____, 5.2, 6.8, _____

⑦ $\frac{2}{3}, \frac{3}{5}, \frac{5}{7}, \frac{7}{11}, \frac{11}{13}, \frac{13}{17}$, _____, _____

⑧ $2\frac{3}{4}, 7\frac{11}{12}, 17, \frac{19}{20}, 29\frac{31}{32}, 41\frac{43}{44}, 53\frac{59}{60}$, _______, _______

7.4 The Fibonacci concept

The Fibonacci sequence begins with 0 and 1, then adds the last two numbers to make the next number. For example, $1+1=2$, $1+2=3$, $2+3=5$, $3+5=8$, and $5+8=13$.

$$0, 1, 1, 2, 3, 5, 8, 13, 21, 34, 55, 89, \ldots$$

Example. Fill in the missing numbers.

(A) $3, 4, 7, 11, 18, 29, \_\_\_\_, \_\_\_\_$ This sequence adds the last two numbers. The next two numbers are $18+29=47$ and $29+47=76$. (This is similar to the Fibonacci sequence.)

Problems. Fill in the missing numbers.

① $9, 12, \_\_\_\_, 33, 54, 87, \_\_\_\_$

② $0.18, 0.32, 0.5, 0.82, 1.32, \_\_\_\_, \_\_\_\_$

③ $0.07, 0.08, 0.15, 0.23, 0.38, 0.61, \_\_\_\_, \_\_\_\_$

④ $\frac{1}{6}, \frac{1}{3}, \frac{1}{2}, \frac{5}{6}, \frac{4}{3}, \_\_\_\_, \_\_\_\_$

⑤ $\frac{3}{4}, 1\frac{1}{4}, 2, 3\frac{1}{4}, 5\frac{1}{4}, 8\frac{1}{2}, \_\_\_\_, \_\_\_\_$

7.5 Additive relationships

In a simple additive relationship, two quantities differ by a fixed amount. For example, in the table below, Kim's age is always 4 more than Emily's age.

Emily	5	6	7	8	9
Kim	9	10	11	12	13

A simple additive relationship can be modeled by a formula of the form $y = x + b$, where x and y are the two quantities and b is a constant. For example, for the table above, if we let E stand for Emily's age and K stand for Kim's age, the data in the table obeys the formula $K = E + 4$. If you plug any of Emily's ages into the formula, you will get Kim's age. For example, when Emily is 5, Kim is $K = 5 + 4 = 9$, and when Emily is 8, Kim is $8 + 4 = 12$. The formula allows you to make a prediction. For example, when Emily is 21, the formula predicts that Kim will be $K = 21 + 4 = 25$.

Example. (A) Fill in the missing numbers.

p	3	6		12	
q	11	14	17		23

Add 3 along the top. On top, $6 + 3 = 9$ and $12 + 3 = 15$. The bottom row is $q = p + 8$. On the bottom, $12 + 8 = 20$.

p	3	6	9	12	15
q	11	14	17	20	23

Problems. Give the formula and fill in the missing numbers.

① Fill in the missing numbers.

A	7	14		28	35
C	16		30		

What is the formula for the table above?

② Fill in the missing numbers.

h	0.375			1.5	1.875
t	2.02	2.395	2.77		

What is the formula for the table above?

③ Fill in the missing numbers.

L	$\frac{1}{3}$		$\frac{2}{3}$	$\frac{5}{6}$	
M	$\frac{1}{4}$	$\frac{5}{12}$			$\frac{11}{12}$

What is the formula for the table above?

Example. (B) Use $z = w + 5$ to complete the table.

w	1	3	5	7	9
z					

Add 5 to w. For example, $1 + 5 = 6$ and $3 + 5 = 8$.

w	1	3	5	7	9
z	6	8	10	12	14

Problems. Complete the table and make the prediction.

④ $V = U + 12$ Predict V when $U = 89$.

U	9	17	25	33	41
V					

⑤ $L = K - 0.27$ Predict L when $K = 5.11$.

K	0.7	1.33	1.96	2.59	3.22
L					

⑥ $Q = P + 2\frac{1}{6}$ Predict Q when $P = 9\frac{1}{3}$.

P	$2\frac{2}{3}$	4	$5\frac{1}{3}$	$6\frac{2}{3}$	8
Q					

7.6 Multiplicative relationships

In a simple multiplicative relationship, two quantities are different by a fixed factor. For example, in the table below, the number of forks is always 3 times the number of spoons.

spoons	1	2	3	4	5
forks	3	6	9	12	15

A simple multiplicative relationship can be modeled by a formula of the form $y = mx$ (which means m times x), where x and y are the two quantities and m is a constant. For example, for the table above, if we let S stand for the number of spoons and F stand for the number of forks, the data in the table obeys the formula $F = 3S$ (meaning 3 times S). If you plug any value of S into the formula, it will give you the corresponding value of F. For example, when $S = 2$, we get $F = 3 \times 2 = 6$.

Example. (A) Fill in the missing numbers.

I	4	8	12		20
V	20	40		80	

Add 4 along the top. On top, $12 + 4 = 16$. The bottom row is $V = 5I$ (meaning 5 times I). On the bottom, $5 \times 12 = 60$ and $5 \times 20 = 100$.

I	4	8	12	16	20
V	20	40	60	80	100

Problems. Give the formula and fill in the missing numbers.

① Fill in the missing numbers.

b	9	18	27		45
e	63			252	

What is the formula for the table above?

② Fill in the missing numbers.

t	0.42	0.84		1.68	
d	2.1		6.3		10.5

What is the formula for the table above?

③ Fill in the missing numbers.

C	$\frac{1}{2}$	$\frac{3}{4}$		$\frac{5}{4}$	$\frac{3}{2}$
Q	3		6		

What is the formula for the table above?

Example. (B) Use $W = 10F$ to complete the table.

F	3	5	7	9	11
W					

Multiply F by 10. For example, $10 \times 3 = 30$.

F	3	5	7	9	11
W	30	50	70	90	110

Problems. Complete the table and make the prediction.

④ $v = 12a$ Predict v when $a = 73$.

a	8	9	10	11	12
v					

⑤ $P = 0.8V$ Predict P when $V = 29$.

V	5	7	9	11	13
P					

⑥ $y = \frac{2}{3}x$ Predict y when $x = 48$.

x	6	8	11	15	20
y					

7.7 Additive graphs

As discussed in Sec. 7.5, an additive relationship has the form $y = x + b$, where x and y are the variables and b is a constant. For example, the table below can be modeled by $y = x + 2$. For example, when $x = 4$, the formula gives $y = 4 + 2 = 6$.

x	0	2	4	6	8
y	2	4	6	8	10

We can plot this table of (x, y) values on a coordinate graph using the method from Sec. 6.10. A coordinate graph for the table above is shown below.

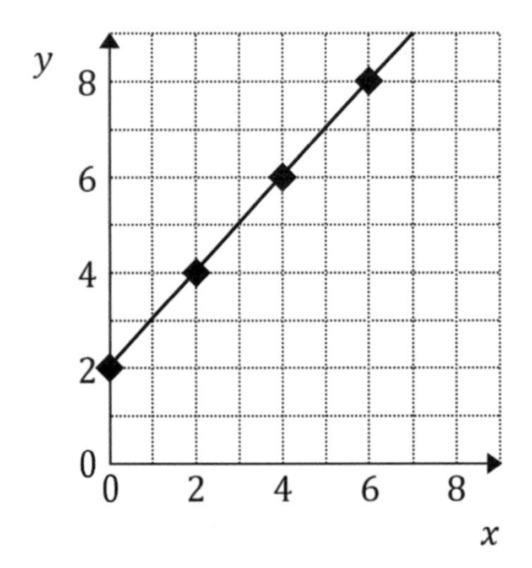

A purely additive graph has a **slope** of one, meaning that the rise equals the run. For example, in the graph above, the line rises 7 units upward (from $y = 2$ to $y = 9$) and runs

7 units across (from $x = 0$ to $x = 7$). The **rise** is how far the line travels upward and the **run** is how far the line travels across horizontally. For a purely additive graph, the rise and run are equal, which means that the slope equals one. On a purely additive graph, the constant b is the y-**intercept**. The y-intercept is the value of y when $x = 0$. For example, the graph of $y = x + 2$ shown on the previous page has a y-intercept of 2: When $x = 0$, $y = 2$. In general, for a purely additive graph, which has an equation of the form $y = x + b$, the y-intercept is b. The line crosses the y-axis at $y = b$.

Example. (A) Make a graph for the data below.

x	0	2	4	6	8
y	1	3	5	7	9

For each value of x, y is one unit larger: $y = x + 1$. Plot each (x, y) pair. The graph has a slope of 1 and a y-intercept of 1.

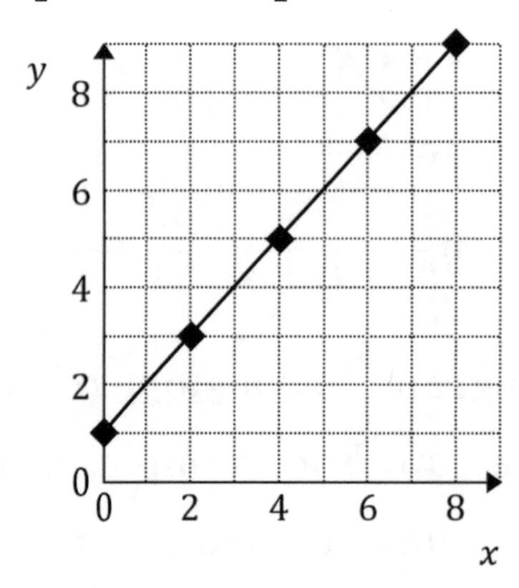

Problems. Draw a coordinate graph for each set of data.

① Draw a coordinate graph for the data below.

x	0	2	4	6	8
y	3	5	7	9	11

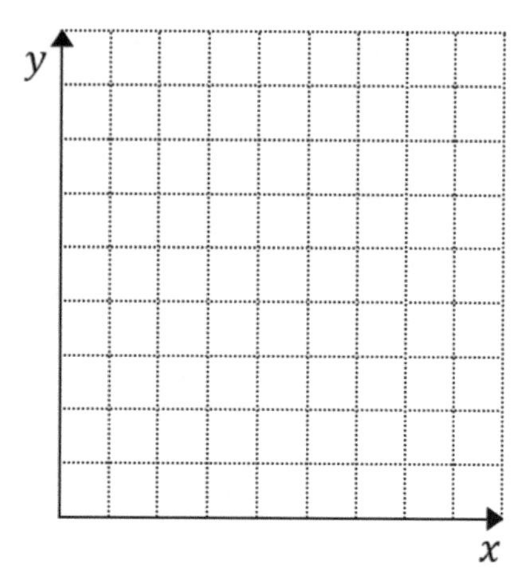

② Draw a coordinate graph for the data below.

x	2	4	6	8	10
y	0	2	4	6	8

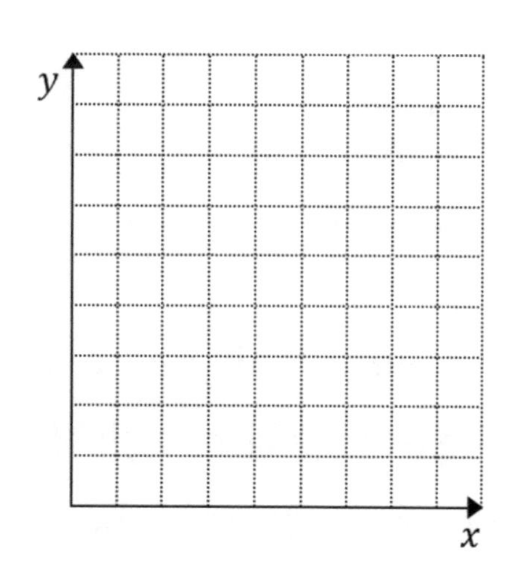

7.8 Multiplicative graphs

As discussed in Sec. 7.6, a multiplicative relationship has the form $y = mx$, where x and y are the variables and m is a constant. For example, the table below can be modeled by $y = 2x$. For example, when $x = 4$, the formula gives $y = 2 \times 4 = 8$.

x	0	2	4	6	8
y	0	4	8	12	16

We can plot this table of (x, y) values on a coordinate graph using the method from Sec. 6.10. A coordinate graph for the table above is shown below.

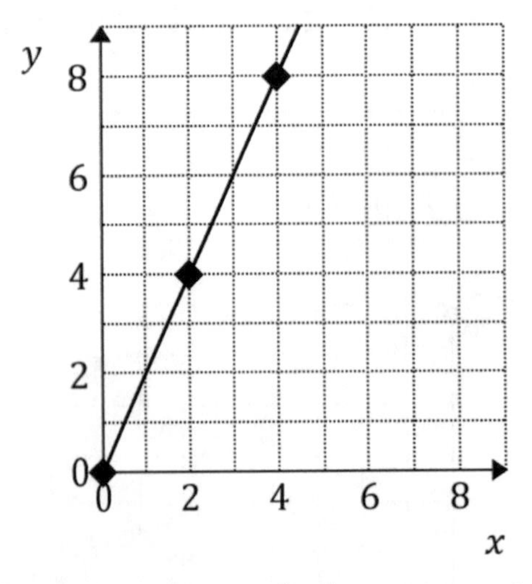

A purely multiplicative graph has a y-intercept of zero, meaning that the line passes through the point (0,0), called the **origin**. The slope of the graph above equals 2, meaning that the line rises two units upward for each unit that it runs across. You can see that the line rises 8 units upward when it goes across 4 units; it passes through the point (4,8). For a purely multiplicative graph, the y-intercept is the constant m. For example, the graph of $y = 2x$ shown on the previous page has a slope of 2. In general, for a purely multiplicative graph, which has an equation of the form $y = mx$, the slope is m.

Example. (A) Make a graph for the data below.

x	0	1	2	3	4
y	0	4	8	12	16

For each value of x, y is 4 times larger: $y = 4x$. Plot each (x, y) pair. The graph has a slope of 4 and a y-intercept of 0.

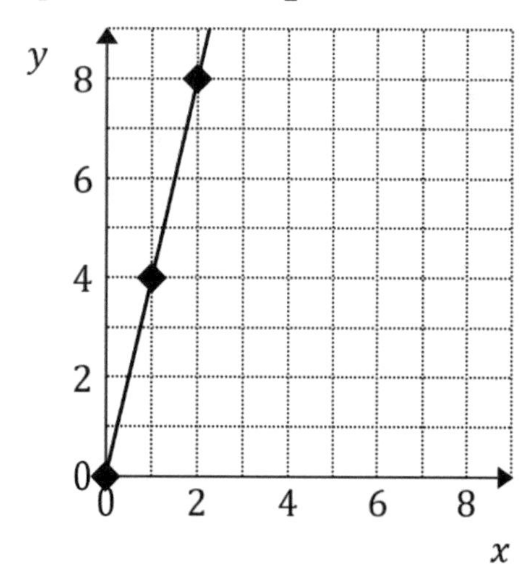

Problems. Draw a coordinate graph for each set of data.

① Draw a coordinate graph for the data below.

x	0	1	2	3	4
y	0	3	6	9	12

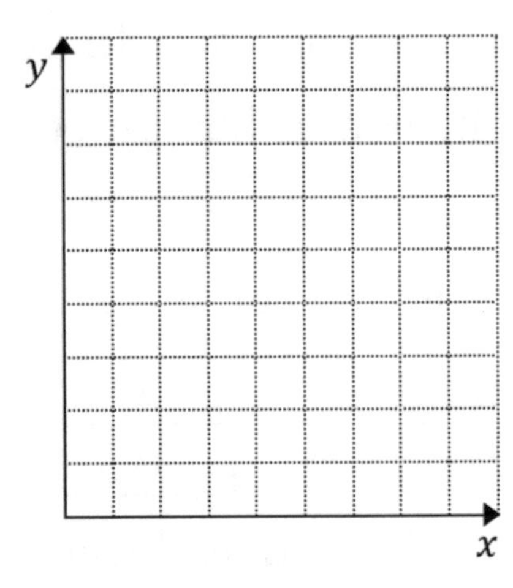

② Draw a coordinate graph for the data below.

x	0	2	4	6	8
y	0	1	2	3	4

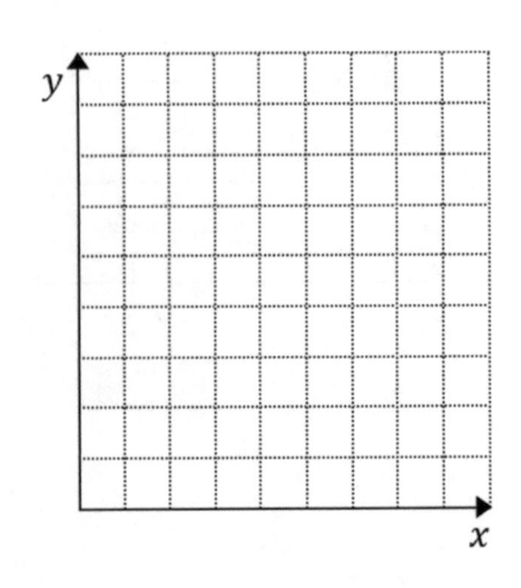

7.9 Word problems

Each word problem in this section involves an additive or a multiplicative relationship.

Examples. **(A)** A box has a weight of 2 pounds. When a boy puts one brick in the box, the total weight is 8 pounds. What will be the total weight if the boy adds a second brick to the box?

One brick weighs $8 - 2 = 6$ pounds. Since the box weighs 2 pounds, the total weight will always be 2 pounds more than the weight of the bricks. This is an additive relationship. Since 2 bricks weigh $6 + 6 = 12$ pounds, the total weight will be $12 + 2 = 14$ pounds.

(B) A man travels 60 yards in 3 minutes. How far will the man walk in 12 minutes?

This is a multiplicative relationship. The distance traveled follows the pattern $60, 120, 180, \boxed{240}, 300, \ldots$ while the time follows the pattern $3, 6, 9, \boxed{12}, 15, \ldots$ The man will travel 240 yards in 12 minutes. Since $60 \div 3 = 20$, the man will travel 20 yards every minute. In 12 minutes, this works out to $12 \times 20 = 240$ yards.

Problems. Solve each word problem.

① A roll of fabric has twice as many stripes as stars. How many stripes are on a section that has 24 stars? How many stripes are on a section that has 72 stars?

② A container already has 3 quarts of water in it. If 6 quarts are added, how much water will the container have? If 6 more quarts are added after that, how much water will the container have?

③ One memory stick can hold up to 16 GB of data. How much data can 7 memory sticks hold? How much data can 25 memory sticks hold?

④ A recipe calls for 3 eggs for every 8 cups of flour. How many eggs are needed for 40 cups of flour? How many eggs are needed for 64 cups of flour?

⑤ A student begins reading a book on page 7. On which page will the student finish reading if the student reads a total of 15 pages? On which page will the student finish reading if the student reads another 15 pages after that?

⑥ A store sells 5 super balls for $3. How much does it cost to buy 20 super balls? How much does it cost to buy 30 super balls?

⑦ An engineer models data with the equation $q = j + 9$. If $j = 7$, what is q? If $j = 76$, what is q?

⑧ Two variables are related by the equation $F = 32G$. If $G = 8$, what is F? If $G = 29$, what is F?

Multiple Choice Questions

① Which is next in the pattern 28, 45, 62, 79, 96?

(A) 103 (B) 112 (C) 113 (D) 124 (E) 175

② Which is next in the pattern 8, 24, 72, 216, 648?

(A) 664 (B) 792 (C) 864 (D) 1824 (E) 1944

③ What is the relationship between w and z below?

(A) additive (B) multiplicative (C) inverse (D) other

w	3	6	9	12	15
z	6	12	18	24	30

④ What is the relationship between D and E below?

(A) additive (B) multiplicative (C) inverse (D) other

D	7	14	21	28	35
E	14	21	28	35	42

⑤ Which models the relationship between B and F below?

(A) $F = 6B$ (B) $B = 6F$ (C) $F = B + 10$ (D) $B = F + 10$

B	2	4	6	8	10
F	12	24	36	48	60

⑥ Which models the relationship between d and h below?

(A) $h = 2d$ (B) $h = 3d$ (C) $d = 3h$ (D) $h = d + 12$

d	6	12	18	24	30
h	18	24	30	36	42

⑦ Given $k = n + 28$, predict k when $n = 45$.

(A) $k = 17$ (B) $k = 28$ (C) $k = 63$ (D) $k = 73$ (E) $k = 1260$

⑧ Given $H = 8G$, predict H when $G = 13$.

(A) $H = 5$ (B) $H = 21$ (C) $H = 84$ (D) $H = 94$ (E) $H = 104$

⑨ In $e = d + 6$, what is the relationship between d and e?

(A) additive (B) multiplicative (C) inverse (D) other

⑩ In $z = 0.3y$, what is the relationship between z and y?

(A) additive (B) multiplicative (C) inverse (D) other

⑪ Which of the following is **NOT** a prime number?

(A) 37 (B) 47 (C) 59 (D) 79 (E) 91

⑫ Which of the following is **NOT** a composite number?

(A) 35 (B) 42 (C) 51 (D) 73 (E) 87

⑬ Which is a pair of prime numbers with a sum that is also a prime number?

(A) 2, 19 (B) 2, 41 (C) 3, 17 (D) 3, 29 (E) 6, 13

⑭ Which number is missing in the table below?

(A) 2.025 (B) 2.125 (C) 2.15 (D) 2.225 (E) 2.25

x	0.425	0.85	1.275	1.7	
y	0.85	1.7	2.55	3.4	4.25

⑮ Which models the relationship between x and y above?

(A) $y = 0.5x$ (B) $y = 2x$ (C) $x = 2y$ (D) $y = x + 0.425$

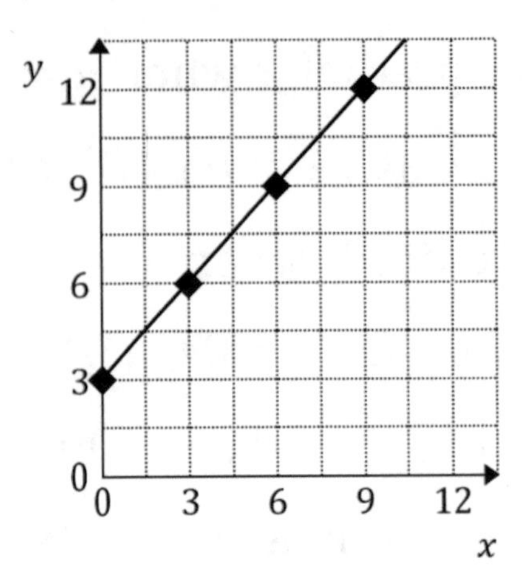

⑯ What is the equation for the graph above?

(A) $y = \frac{x}{2}$ (B) $y = 2x$ (C) $x = 2y$ (D) $y = x + 3$

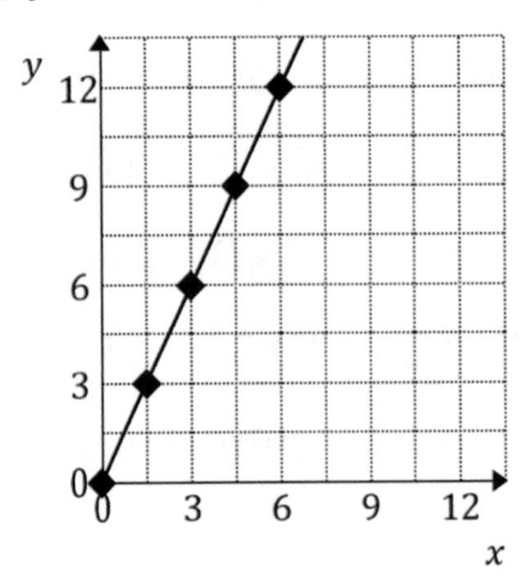

⑰ What is the equation for the graph above?

(A) $y = \frac{x}{2}$ (B) $y = 2x$ (C) $x = 2y$ (D) $y = x + 3$

⑱ The relationship in the graph for Exercise 17 is _______.

(A) additive (B) multiplicative (C) inverse (D) other

⑲ If 3 slices of pizza cost \$8, how much does it cost to buy 42 slices of pizza?

(A) \$24 (B) \$112 (C) \$122 (D) \$126 (E) \$336

-8-

VARIABLES

8.1 Variables and constants

In math, a letter like x or y is called a **variable**. The reason that it is called a variable is that its value may change from one problem to another. For example, $x = 3$ satisfies the equation $x + 2 = 5$ while $x = 6$ satisfies the equation $x + 3 = 9$. The value of x is different in each equation. A variable is an unknown quantity that we are solving for. A number like 8 or 4.7 is called a **constant**. A constant that multiplies a variable is called a **coefficient**, like the 2 in the expression $2x$ (which means 2 times x).

Example. (A) In the equation $3y + 2 = 8$, y is the variable, 3 is the coefficient, and 3, 2, and 8 are constants.

Problems. Identify the variable, coefficient, and constants. You don't need to solve the equation.

① $4y - 3 = 5$

② $2t + 9 = 5t$

③ $7 + 6x = 25$

8.2 Balancing additive equations

If an equation contains a variable, we can visualize it like balancing a scale. The equal sign (=) indicates that the scale is balanced, meaning that the left-hand side and the right-hand side of equation are equal. For example, consider the equation $x + 2 = 6$. To draw a scale for this equation, put x and 2 blocks on the left, and put 6 blocks on the right. Since the scale is balanced, both sides must have the equivalent of 6 blocks. In order for the left side to have 6 blocks, the value of x in this equation must be 4 blocks. The equation $x + 2 = 6$ is solved by $x = 4$ because $4 + 2 = 6$.

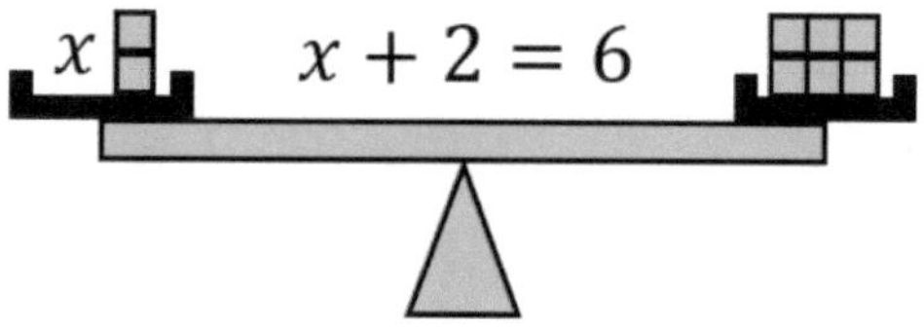

$x + 2 = 6$

Example. (A) Draw a scale for $5 + x = 7$ and determine the value of x.

Draw a scale with 5 and x on the left, and 7 on the right.

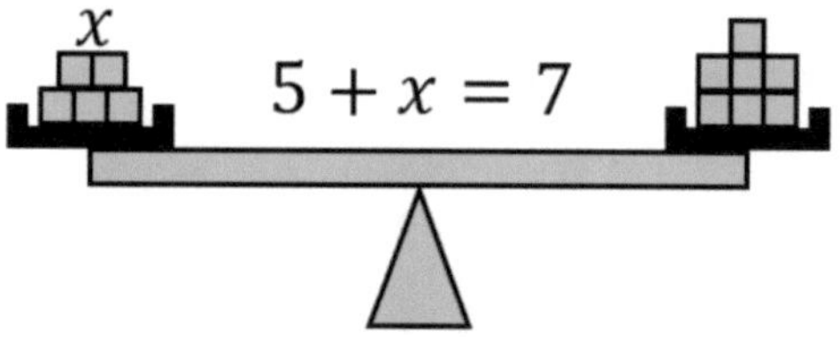

$5 + x = 7$

In order for the left side to have 7 blocks, the value of x in this equation must be 2 because $5 + 2 = 7$.

Problems. Draw a scale to represent the balanced equation and determine the value of x.

① $x + 4 = 9$

② $3 + x = 10$

③ $16 = 7 + x$

④ $14 = x + 6$

⑤ $4 + x = 10$

⑥ $8 = 7 + x$

⑦ $17 = x + 8$

⑧ $x + 9 = 12$

8.3 Equations with addition

If a constant is added to the variable, subtract the constant from both sides of the equation. For example, to solve for x in the equation $x + 3 = 5$, subtract 3 from both sides of the equation to get $x = 5 - 3$, which simplifies to $x = 2$. Check the answer by plugging 2 in for x in the original equation: Since $2 + 3 = 5$, this shows that $x = 2$ solves $x + 3 = 5$.

Examples. Solve for x in each equation.

(A) $x + 4 = 7$

Subtract 4 from both sides: $x = 7 - 4 = 3$.

Check the answer by plugging 4 in for x: Since $3 + 4 = 7$, the answer checks out.

(B) $4 = x + 3$

Subtract 3 from both sides: $4 - 3 = 1 = x$.

Check the answer by plugging 1 in for x: Since $4 = 1 + 3$, the answer checks out.

Problems. Solve for x in each equation. Check the answer by plugging it into the original equation.

① $x + 6 = 12$

② $3 + x = 8$

③ $15 = 7 + x$

④ $x + 9 = 13$

⑤ $3 + x = 5$

⑥ $11 = 5 + x$

⑦ $14 = x + 7$

⑧ $x + 7 = 12$

⑨ $4 + x = 4$

⑩ $18 = 9 + x$

⑪ $10 = 6 + x$

⑫ $8 + x = 16$

⑬ $x + 9 = 16$

⑭ $12 = 4 + x$

8.4 Equations with subtraction

If a constant is subtracted from a variable, add the constant to both sides of the equation. For example, to solve for x in the equation $x - 2 = 4$, add 2 to both sides of the equation to get $x = 4 + 2$, which simplifies to $x = 6$. Check the answer by plugging 6 in for x in the original equation: Since $6 - 2 = 4$, this shows that $x = 6$ solves $x - 2 = 4$.

Examples. Solve for x in each equation.

(A) $x - 1 = 4$

Add 1 to both sides: $x = 4 + 1 = 5$.

Check the answer by plugging 5 in for x: Since $5 - 1 = 4$, the answer checks out.

(B) $3 = x - 3$

Add 3 to both sides: $3 + 3 = 6 = x$.

Check the answer by plugging 6 in for x: Since $3 = 6 - 3$, the answer checks out.

Problems. Solve for x in each equation. Check the answer by plugging it into the original equation.

① $x - 3 = 6$ ② $7 = x - 5$

③ $4 = x - 4$

④ $x - 8 = 9$

⑤ $x - 6 = 5$

⑥ $8 = x - 5$

⑦ $7 = x - 8$

⑧ $x - 9 = 9$

⑨ $x - 6 = 0$

⑩ $3 = x - 5$

⑪ $5 = x - 9$

⑫ $x - 4 = 8$

⑬ $x - 1 = 1$

⑭ $6 = x - 8$

8.5 Addition and subtraction equations

Problems. Solve for x in each equation. Check the answer by plugging it into the original equation.

① $x + 5 = 10$

② $7 = x - 9$

③ $x - 4 = 6$

④ $x + 8 = 14$

⑤ $7 = x - 6$

⑥ $11 = 7 + x$

⑦ $13 = 5 + x$

⑧ $x - 7 = 8$

⑨ $x + 7 = 16$

⑩ $9 = x - 2$

8.6 Strip diagrams

In Sec. 8.2, we saw that we can visualize an equation with a variable using a balanced scale. In this section, we will see that another way to visualize an equation with addition or subtraction is to use a **strip diagram**. Follow these steps to draw a strip diagram for a simple addition or subtraction equation that has one variable and two constants:

- Identify the largest value. Draw one long strip for this value.
- Draw two strips joined together to represent the other two values. The combined lengths of these two strips must match the length of the strip from the first step.

For example, to draw a strip diagram for $15 + x = 20$, draw one long strip for 20 and draw two shorter strips for x and 15. The strips for x and 15 join together to form 20.

20	
15	x

Since 15 and x add together to form 20, the value of x must be 5. Check the answer by plugging 5 in for x in the equation. Since $15 + 5 = 20$, the answer checks out.

Examples. **(A)** Draw a strip diagram for $12 + x = 24$.

The largest value is 24. The values 12 and x add up to 24. Draw a long strip for 24. Draw two strips joined together for 12 and x.

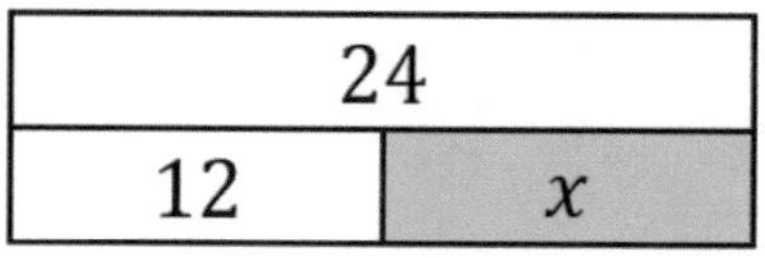

24	
12	x

Since 12 and x add together to form 24, the value of x must be 12. Check the answer by plugging 12 in for x in the equation. Since $12 + 12 = 24$, the answer checks out.

(B) Draw a strip diagram for $x - 10 = 15$.

The largest value is x. Why? Because 10 is subtracted from x to make 15. This shows that x must be larger than 15. Draw a long strip for x. Draw two strips joined together for 10 and 15.

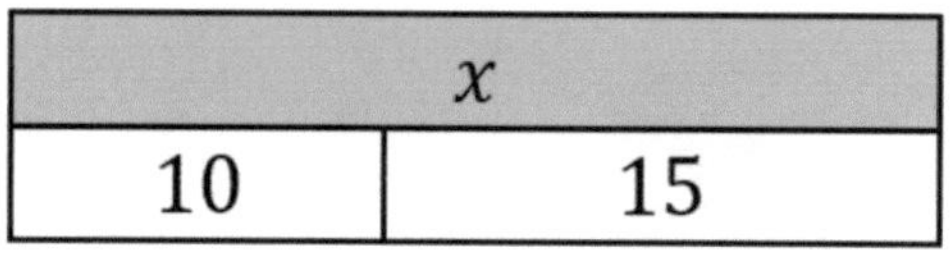

x	
10	15

Since 10 and 15 add together to form x, the value of x must be 25. Check the answer by plugging 25 in for x in the equation. Since $25 - 10 = 15$, the answer checks out.

Problems. Draw a strip diagram to represent the equation and determine the value of x.

① $x + 30 = 45$

② $x - 18 = 54$

③ $7 + x = 21$

④ $16 = 48 - x$

⑤ $100 - x = 64$

⑥ $x + 2.5 = 4.2$

⑦ $\frac{1}{3} = x - \frac{1}{6}$

⑧ $75 - x = 25$

8.7 Equations with multiplication

It is common to avoid the standard times symbol (×) when multiplying with variables. This helps to avoid confusion with the symbol x, which is commonly used as a variable. For example, $2x$ means 2 times x. If we wrote this as $2 \times x$, the two crosses (× and x) might seem confusing, especially when writing with your hand.

If a constant multiplies the variable, divide both sides of the equation by the constant. (A constant that multiplies a variable is called a **coefficient**. We are dividing both sides of the equation by the coefficient.) For example, to solve for x in the equation $2x = 8$, divide each side of the equation by 2 to get $x = 8 \div 2$, which simplifies to $x = 4$. Check the answer by plugging 4 in for x in the original equation: Since $2 \times 4 = 8$, this shows that $x = 4$ solves $2x = 8$.

Example. (A) Solve for x in $3x = 6$.

Divide by 3 on both sides: $x = 6 \div 3 = 2$.

Check the answer: Since $3 \times 2 = 6$, the answer checks out.

Problems. Solve for x in each equation. Check the answer by plugging it into the original equation.

① $4x = 24$

② $7x = 14$

③ $15 = 5x$

④ $5x = 35$

⑤ $3x = 12$

⑥ $48 = 6x$

⑦ $81 = 9x$

⑧ $8x = 56$

⑨ $5x = 5$

⑩ $63 = 7x$

8.8 Equations with division

It is common to use a fraction to represent division when working with variables. For example, $\frac{x}{2}$ means x divided by 2. This is usually done instead of using the division symbol ($\div$). We can do this with numbers, too. For example, $\frac{18}{3} = 6$.

If a variable is divided by a constant, multiply both sides of the equation by the constant. For example, to solve for x in the equation $\frac{x}{3} = 5$, multiply both sides of the equation by 3 to get $x = 5 \times 3$, which simplifies to $x = 15$. Check the answer by plugging 15 in for x in the original equation: Since $\frac{15}{3} = 15 \div 3 = 5$, this shows that $x = 15$ solves $\frac{x}{3} = 5$.

Example. (A) Solve for x in $\frac{x}{2} = 6$.

Multiply by 2 on both sides: $x = 2 \times 6 = 12$.

Check the answer: Since $\frac{12}{2} = 12 \div 2 = 6$, the answer checks out.

Problems. Solve for x in each equation. Check the answer by plugging it into the original equation.

① $\frac{x}{4} = 7$

② $\frac{x}{8} = 4$

③ $9 = \frac{x}{2}$

④ $\frac{x}{5} = 5$

⑤ $\frac{x}{7} = 6$

⑥ $\frac{x}{9} = 3$

⑦ $4 = \frac{x}{3}$

⑧ $3 = \frac{x}{6}$

⑨ $\frac{x}{9} = 6$

⑩ $9 = \frac{x}{8}$

8.9 One-step equations

One-step equations are simple enough to solve in a single step. To solve a one-step equation, examine what is being done to the variable and then do the **opposite**.

- Subtraction is the opposite of addition.
- Division is the opposite of multiplication.

For example, in $x + 5 = 12$, since 5 is *added* to x we should *subtract* 5 from both sides to get $x = 12 - 5 = 7$. Similarly, in $4x = 8$, since 4 is *multiplying* x, we should *divide* by 4 on both sides to get $x = \frac{8}{4} = 8 \div 4 = 2$.

Examples. Solve for x in each equation.

(A) $x - 4 = 2$

Do the opposite: Since 4 is *subtracted* from x, *add* 4 to both sides: $x = 2 + 4 = 6$. Check the answer by plugging 6 in for x: Since $6 - 4 = 2$, the answer checks out.

(B) $\frac{x}{3} = 3$

Do the opposite: Since x is *divided* by 3, *multiply* by 3 on both sides: $x = 3 \times 3 = 9$. Check the answer by plugging 9 in for x: Since $\frac{9}{3} = 9 \div 3 = 3$, the answer checks out.

Problems. Solve for x in each equation. Check the answer by plugging it into the original equation.

① $x - 27 = 48$

② $4x = 148$

③ $\frac{x}{17} = 26$

④ $8.3 = x + 3.5$

⑤ $13x = 156$

⑥ $\frac{2}{3} = x - \frac{1}{12}$

⑦ $9x = \frac{7}{4}$

⑧ $\frac{x}{7} = \frac{3}{14}$

⑨ $x + 276 = 431$

⑩ $3.2 = 4x$

8.10 Two-step equations

Two-step equations can be solved by **isolating the unknown**:

- First move the constant term to the other side by doing the opposite. For example, in $2x + 4 = 10$, since 4 is *added*, *subtract* 4 from both sides to get $2x = 10 - 4$, and in $3x - 1 = 8$, since 1 is *subtracted*, *add* 1 to both sides to get $3x = 8 + 1$.
- Move the remaining constant to the other side by doing the opposite. This **isolates** the unknown. For example, in $2x = 6$, since 2 is *multiplying*, *divide* by 2 on both sides to get $x = 6 \div 2$, and in $\frac{x}{4} = 12$, since x is *divided* by 4, *multiply* by 4 on both sides to get $x = 4 \times 12$.

Note that we perform the addition or subtraction step first and the multiplication or division step last.

Examples. Solve for x in each equation.

(A) $5x - 3 = 7$

Since 3 is *subtracted* from $5x$, *add* 3 to both sides: $5x = 7 + 3 = 10$. Since x is *multiplied* by 5, *divide* by 5 on both sides: $x = 10 \div 5 = 2$. Check the answer by plugging 2 in for x: Since $5x - 3 = 5 \times 2 - 3 = 10 - 3 = 7$, the answer checks out.

(B) $\frac{x}{2} + 1 = 4$

Since 1 is *added* to $\frac{x}{2}$, *subtract* 1 from both sides: $\frac{x}{2} = 4 - 1 = 3$. Since x is *divided* by 2, *multiply* by 2 on both sides: $x = 3 \times 2 = 6$. Check the answer by plugging 6 in for x: Since $\frac{6}{2} + 1 = 6 \div 2 + 1 = 3 + 1 = 4$, the answer checks out.

Problems. Solve for x in each equation. Check the answer by plugging it into the original equation.

① $3x + 7 = 25$

② $8x - 18 = 22$

③ $\frac{x}{3} - 4 = 5$

④ $60 = 6x + 24$

⑤ $\frac{x}{7} + 2 = 9$

⑥ $21 = 4x - 15$

⑦ $61 = 19 + 7x$

⑧ $2 = \frac{x}{6} - 2$

⑨ $\frac{x}{8} - 1 = 7$

⑩ $100 = 28 + 9x$

8.11 Modeling word problems

One way to solve a word problem is to write an equation with a variable and then solve for the variable in the equation.

Examples. (A) After Angela ate 5 strawberries, there were 7 strawberries left. How many strawberries were there in the beginning?

- Let x represent the original number of strawberries.
- Angela reduced x strawberries by 5, after which there were 7 strawberries left: $x - 5 = 7$.
- Solve the equation: $x = 7 + 5 = 12$.
- Check the answer: $12 - 5 = 7$.

(B) Maria has some hairbands. Maria divides the hairbands into 6 equal piles. Each pile has 3 hairbands. How many hairbands does Maria have?

- Let x represent the number of hairbands.
- Maria divided x hairbands into 6 piles, and each pile has 3 hairbands: $\frac{x}{6} = 3$.
- Solve the equation: $x = 6 \times 3 = 18$.
- Check the answer: $\frac{18}{6} = 18 \div 6 = 3$.

Problems. Use an equation to solve each problem.

① After Cynthia spent $18, she had $24 remaining. How much money did Cynthia have originally?

② Kyle bought some trading cards. Each trading card costs $6. The total cost was $102. How many trading cards did Kyle buy? (There is no sales tax.)

③ Rebecca originally had 7 hats. After receiving new hats, Rebecca had a total of 11 hats. How many hats did Rebecca receive?

④ Roger has a box of marbles. Roger divided the marbles into 9 piles. Each pile has 13 marbles. How many marbles are there all together?

⑤ Linda has a photo album. When Linda added 8 pictures to her photo album, there were a total of 27 pictures in the album. How many pictures were in the photo album before she added the pictures?

⑥ Paul has a super stretchy cable. When Paul stretches the cable until it is 6 times its original length, the cable is 192 inches long. What is the original length of the cable?

⑦ William has some seeds. William plants the seeds in 12 rows. Each row has 14 seeds. How many seeds are there all together?

⑧ Martha has some flowers. After Martha gave 7 flowers to her friends, she had 17 flowers left. How many flowers did Martha have originally?

Multiple Choice Questions

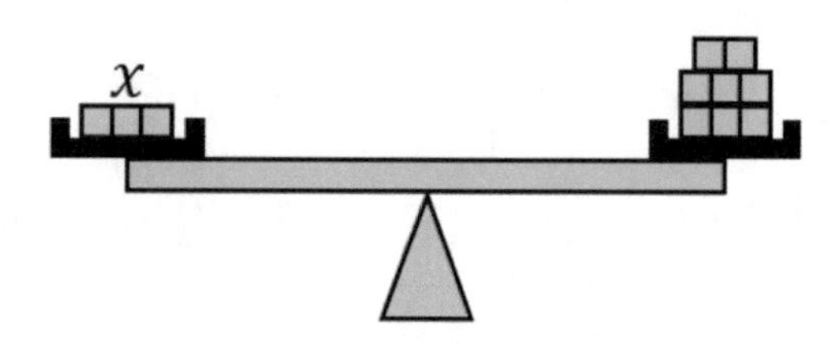

① What is the value of x on the balanced scale above?

(A) 3 (B) 4 (C) 5 (D) 7 (E) 8

② Which value of x solves $x + 25 = 50$?

(A) 0.5 (B) 2 (C) 25 (D) 75 (E) 1250

③ Which value of x solves $x - 16 = 48$?

(A) $\frac{1}{3}$ (B) 3 (C) 32 (D) 64 (E) 768

④ Which value of x solves $9x = 72$?

(A) $\frac{1}{8}$ (B) 8 (C) 63 (D) 81 (E) 648

⑤ Which value of x solves $\frac{x}{4} = 24$?

(A) $\frac{1}{6}$ (B) 6 (C) 20 (D) 28 (E) 96

<table>
<tr><td colspan="2">20</td></tr>
<tr><td>x</td><td>12</td></tr>
</table>

⑥ Which equation does the above strip diagram represent?

(A) $12x = 20$ (B) $x + 12 = 20$ (C) $x - 12 = 20$

(D) $\frac{x}{12} = 20$ (E) $20 + x = 12$

⑦ Which value of x solves $6x - 30 = 54$?

(A) 4 (B) 9 (C) 14 (D) 84 (E) 504

⑧ Which value of x solves $\frac{x}{4} + 12 = 20$?

(A) 2 (B) 8 (C) 16 (D) 32 (E) 128

⑨ Sylvia has a basket of eggs. After Sylvia gave 18 eggs to her friends, she had 36 eggs left. Which equation can be used to model this problem?

(A) $x + 18 = 36$ (B) $x - 18 = 36$ (C) $36 - x = 18$

(D) $\frac{x}{18} = 36$ (E) $18x = 36$

⑩ In Exercise 9, how many eggs did Sylvia have originally?

(A) 2 (B) 18 (C) 54 (D) 72 (E) 648

⑪ Alan divides his stickers into 8 piles. Each pile has 24 stickers. Which equation can be used to model this problem?

(A) $x + 8 = 24$ (B) $x - 8 = 24$ (C) $24 - x = 8$

(D) $\frac{x}{8} = 24$ (E) $8x = 24$

⑫ In Exercise 11, how many stickers does Alan have all together?

(A) 3 (B) 16 (C) 32 (D) 162 (E) 192

-9-

MEASUREMENT AND GEOMETRY

9.1 Customary measurements

Customary measures of **length** include inches, feet, yards, and miles. There are 12 **inches** in one **foot**, 3 feet in one **yard**, and 1760 yards in one **mile** (or 5280 feet in one mile). The abbreviations for these units are in. for inch, ft for foot, yd for yard, and mi for mile. Length is commonly measured with one of the following measuring devices:

- Most **rulers** are 1 foot long. Rulers are suitable for measuring short distances in inches.
- A **yardstick** is 1 yard long. A yardstick is well-suited for measurements between 1 and 3 feet.
- A **tape measure** can measure several feet.
- A **trundle wheel** rolls along the ground and is suitable for measurements along which a person can walk.
- An **odometer** on a car is well-suited for long distances that a person can drive.

Customary measures of **capacity** include fluid ounces, cups, pints, quarts, and gallons. In the United States, there are 8 **fluid ounces** in one **cup** (or 16 fluid ounces in one pint), 2 cups in one **pint**, 2 pints in one **quart**, and 4 quarts in one

gallon. The abbreviations for these units are fl oz for fluid ounce, c for cup, pt for pint, qt for quart, and gal for gallon. Capacity is commonly measured with one of the following measuring devices:

- **Measuring spoons** often come in a set with different sizes and make measurements in **teaspoons** or **tablespoons**. The tablespoon is larger: 1 tablespoon equates to 3 teaspoons. One fluid ounce equals 2 tablespoons or 6 teaspoons.
- **Measuring cups** often come in a set with different sizes, such as from $\frac{1}{4}$ cup to 1 cup (or larger). The size of the measuring cup is well-suited for measurements that are smaller than its size (but not way smaller). For example, a $\frac{1}{2}$-cup measuring cup can measure up to $\frac{1}{2}$ of a cup and down to about $\frac{1}{16}$ of a cup (depending on how it is graduated).
- A variety of containers, such as jugs or cartons, may hold larger capacities.

Customary measures of **weight** include ounces, pounds, and tons. In the United States, there are 16 **ounces** in one **pound**, and 2000 pounds in one **ton**. The abbreviations for these units are oz for ounce, lb for pound, and T for ton. Weight is commonly measured with one of the following measuring devices:

- Many common **balances** and **scales** can measure small objects in pounds and ounces.
- Certain special scales, like a truck scale, can measure very heavy objects.

Examples. **(A)** Which measuring device is best suited for measuring the height of a coffee table?

Since most coffee tables are taller than a foot, but smaller than a yard, a yardstick would probably be the best.

(B) Which measuring device is best suited for measuring the weight of a person?

There are a variety of scales suitable for this. A scale that a patient stands on in a doctor's office would work. A bathroom scale would work, too.

(C) Which measuring device is best suited for measuring the amount of milk in a small bowl?

A measuring cup that has more capacity (but not way more) than the bowl would work well.

Problems. Which measuring device is best suited for each measurement? Explain the reasoning behind your answer.

① the length of a child's tennis racket

② the width of a piece of notebook paper

③ the distance around a person's waist

④ the height of a doorway

⑤ the length of a basketball court

⑥ the distance between two cities

⑦ the weight of a pocket calculator

⑧ the amount of milk needed to make a few pancakes

Example. (D) Which customary unit is best suited for the length of a tee shot in golf?
It would be appropriate to measure this in yards because a typical tee shot in golf travels two to three hundred yards (which isn't long enough to use miles).

Problems. For each measurement, indicate the customary unit that is best to use.

⑨ a person's height

⑩ a person's weight

⑪ the amount of gasoline in the tank of a car

⑫ the amount of medicine administered to a patient

⑬ the distance a person drives from home to work

⑭ the length of the sleeve of a coat

9.2 Metric measurements

The metric system measures length in **meters**, capacity in **liters**, and mass in **grams**. (Weight and mass are somewhat different, but similar in that they are proportional. If an object has more mass, it will also have more weight. You should learn about the distinction in a *science* class.) The abbreviations for these units are m for meter, L for liter, and g for gram. If the **base unit** (the meter, the liter, or the gram) is smaller or larger than needed to make a particular measurement, we put a **prefix** in front of the unit. Common metric prefixes include **kilo** (k) for 1000, **hecto** (h) for 100, **deka** (dk) for 10, **deci** (d) for $\frac{1}{10}$, **centi** (c) for $\frac{1}{100}$, and **milli** (m) for $\frac{1}{1000}$. For example, cm is short for centimeter and means $\frac{1}{100}$ of a meter. There are:

- 100 centimeters in one meter
- 1000 millimeters in one meter
- 10 millimeters in one centimeter
- 1000 meters in one kilometer

There are similar relationships for liters and grams, such as 1000 milliliters in one liter.

Examples. Fill in each blank.

(A) 1 kg means ___________ g. The answer is 1000.

(B) 1 cm means ___________ m. The answer is $\frac{1}{100}$ or 0.01.

Problems. Fill in each blank.

① 1 cg means ___________ g.

② 1 mL means ___________ L.

③ 1 km means ___________ m.

④ 1 mg means ___________ g.

⑤ 1 cL means ___________ L.

⑥ 1 dkg means ___________ g.

⑦ 1 dL means ___________ L.

⑧ 1 hm means ___________ m.

9.3 Customary unit conversions

Suppose that we wish to convert 6 ft to inches or we wish to convert 6 ft to yards. There is more than one way to do this. We will illustrate this with a few different methods.

All of the methods use the following conversion factors:

$$1 \text{ ft} = 12 \text{ in.}$$

$$1 \text{ yd} = 3 \text{ ft}$$

One way is to determine if the new measure uses smaller or larger units than the original measure:

- If the new unit is smaller than the old unit, multiply. Since an inch is smaller than a foot, to convert 6 ft to inches, we multiply: $6 \text{ ft} = 6 \times 12 \text{ in.} = 72 \text{ in.}$
- If the new unit is larger than the old unit, divide. Since a yard is larger than a foot, to convert 6 ft to yards, we divide: $6 \text{ ft} = 6 \div 3 \text{ yd} = 2 \text{ yd}$.

Another way is to make a fraction equal to one using the conversion factor. Write the old unit in the denominator and the new unit in the numerator. The old unit will cancel out. See the examples on the next page.

- To convert 6 ft to inches, multiply by $\frac{12 \text{ in.}}{1 \text{ ft}}$. The feet cancel out. The math is 6 times 12 divided by 1.

$$6 \text{ ft} \times \frac{12 \text{ in.}}{1 \text{ ft}} = 72 \text{ in.}$$

- To convert 6 ft to yards, multiply by $\frac{1 \text{ yd}}{3 \text{ ft}}$. The feet cancel out. The math is 6 times 1 divided by 3.

$$6 \text{ ft} \times \frac{1 \text{ yd}}{3 \text{ ft}} = 2 \text{ yd}$$

Yet another way is to draw strip diagrams, as shown below.

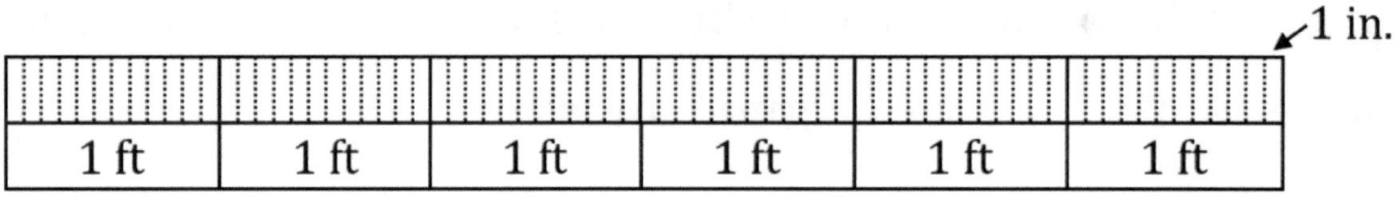

1 in.

1 ft	1 ft	1 ft	1 ft	1 ft	1 ft

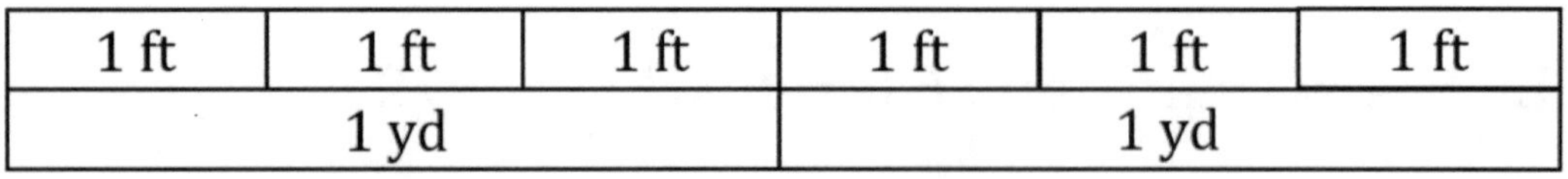

1 ft	1 ft	1 ft	1 ft	1 ft	1 ft
1 yd			1 yd		

The top strip diagrams above show visually that 6 ft equates to 72 in. by dividing each foot into 12 inches. The bottom strip diagrams above show visually that 6 ft equates to 2 yards by grouping every set of 3 feet into a yard.

The conversion factors needed to convert between common customary units are organized in the following tables.

Note: Beware that online conversion calculators may use metric values for pints, cups, or tons unless you choose "US liquid pint," "US cup," or "US ton."

Customary Length Conversions	
1 ft = 12 in.	1 yd = 3 ft
1 mi = 5280 ft	1 mi = 1760 yd

Customary Capacity Conversions	
1 c = 8 fl oz	1 pt = 2 c
1 qt = 2 pt	1 pt = 16 fl oz
1 gal = 4 qt	1 gal = 8 pt

Customary Weight Conversions	
1 lb = 16 oz	1 T = 2000 lb

Note: A "US liquid pint" is different from a "metric pint," a "US cup" is different from a "metric cup," and a "US ton" is different from a "metric ton."

Examples. (A) Convert 24 in. to feet.

Since a foot is larger than an inch, there will be fewer feet than inches. The answer will have a smaller number than 24. Use the conversion factor 1 ft = 12 in.

$$24 \text{ in.} \times \frac{1 \text{ ft}}{12 \text{ in.}} = 24 \div 12 \text{ ft} = \boxed{2 \text{ ft}}$$

(B) Convert 3 gal to quarts.

Since a quart is smaller than a gallon, there will be more quarts than gallons. The answer will have a larger number than 3. Use the conversion factor 1 gal = 4 qt.

$$3 \text{ gal} \times \frac{4 \text{ qt}}{1 \text{ gal}} = 3 \times 4 \text{ qt} = \boxed{12 \text{ qt}}$$

Problems. Perform the indicated conversion.

① Convert 36 feet to inches.

② Convert 21 feet to yards.

③ Convert 15 yards to feet.

④ Convert 168 in. to feet.

⑤ Convert 4 yards to inches.

⑥ Convert 32 pt to fluid ounces.

⑦ Convert 48 qt to gallons.

⑧ Convert 16 gal to pints.

⑨ Convert 48 lb to ounces.

⑩ Convert 80 oz to pounds.

9.4 Metric unit conversions

Conversions between metric units (like from kg to g) involve powers of ten. To convert from one metric unit to another, follow these steps:

- Find the "from" prefix and "to" prefix in the table. If there is no prefix (as in g or L), this column is "none."
- Start at the "from" prefix and go to the "to" prefix.
- If you traveled to the **right**, multiply by 10 for each step. For example, going from hm to dm is 3 steps to the right, so we would **multiply** by $10 \times 10 \times 10 = 1000$.
- If you traveled to the **left**, divide by 10 for each step. For example, going from mg to dg is 2 steps to the left, so we would **divide** by $10 \times 10 = 100$.

Metric Prefixes						
→ Going right → Multiply by 10 for each step →						
k = 1000	h = 100	dk = 10	none = 1	d = 0.1	c = 0.01	m = 0.001
← Divide by 10 for each step ← Going left ←						

It may help to review Sec.'s 4.4 and 4.11 regarding how to multiply and divide decimals by powers of ten.

Examples. (A) Convert 3.4 m to cm.

"From" is none (since there isn't a prefix before the m in 3.4 m) and "to" is c. Going from none to c on the table is 2 steps to the **right**, so we **multiply** by $10 \times 10 = 100$. When we multiply by a power of ten, we move the decimal point one place to the right for each zero. In this case, we move the decimal point 2 places to the right. Note that 3.4 is equivalent to 3.40 such that moving the decimal point 2 places to the right makes 340.

$$3.4 \text{ m} = 3.4 \times 100 \text{ cm} = 340 \text{ cm}$$

(B) Convert 6.1 dL to hL.

"From" is d and "to" is h. Going from d to h on the table is 3 steps to the **left**, so we **divide** by $10 \times 10 \times 10 = 1000$. When we divide by a power of ten, we move the decimal point one place to the left for each zero. In this case, we move the decimal point 3 places to the left. This turns 6.1 into 0.0061.

$$6.1 \text{ dL} = 6.1 \div 1000 \text{ hL} = 0.0061 \text{ hL}$$

Problems. Perform the indicated conversion.

① Convert 2.7 kg to g.

② Convert 54 mL to L.

③ Convert 3.15 cm to m.

④ Convert 0.42 dg to dkg.

⑤ Convert 4 hL to cL.

⑥ Convert 9.6 kg to mg.

9.5 Comparing measurements

To compare two measurements in different units, convert one of the measurements so that both measurements are in the same units. For example, to compare 3417 lb with 2 T, first convert 2 T to pounds (using the conversion factor 1 T = 2000 lb):

$$2\text{ T} \times \frac{2000\text{ lb}}{1\text{ T}} = 2 \times 2000\text{ lb} = 4000\text{ lb}$$

Since 3417 lb < 4000 lb, it follows that 3417 lb < 2 T.

Examples. (A) Compare 7 ft to 2 yd.

Since a yard is larger than a foot, we'll convert 2 yd to feet:

$$2\text{ yd} \times \frac{3\text{ ft}}{1\text{ yd}} = 2 \times 3\text{ ft} = 6\text{ ft}$$

Since 7 ft > 6 ft, it follows that 7 ft > 2 yd.

(B) Compare 47 mm to 5.3 cm.

Since a cm is larger than a mm, we'll convert 5.3 cm to mm. Going from cm to mm is 1 step to the right, so we multiply by 10:

$$5.3\text{ cm} = 5.3 \times 10\text{ mm} = 53\text{ mm}$$

Since 47 mm < 53 mm, it follows that 47 mm < 5.3 cm.

Problems. Write >, <, or = between the measurements.

① 40 in. 5 ft

② 0.4 L 300 mL

③ 2.4 kg 2000 g

④ 8 qt 2 gal

⑤ 3.5 T 7500 lb

⑥ 12 hm 0.9 km

⑦ 7 lb 100 oz

⑧ 7 dg 0.8 g

⑨ 1.4 cL 14 mL

⑩ 75 in. 3 yd

9.6 Word problems with measures

The word problems in this section involve measurements with different units. It may help to review arithmetic with decimals in Chapter 4.

Example. (A) Deana has 5 glasses that each contain 0.4 qt of lemonade. Deana pours all of the lemonade into a single pitcher. How many gallons of lemonade are in the pitcher? First multiply 5 by 0.4 qt to determine the total capacity of lemonade in quarts: 5×0.4 qt = 2 qt. Now convert 2 qt to gallons:

$$2 \text{ qt} \times \frac{1 \text{ gal}}{4 \text{ qt}} = 2 \div 4 \text{ gal} = \boxed{0.5 \text{ gal}}$$

Deana has a total of 0.5 gal of lemonade. Note: $\frac{2}{4} = \frac{1}{2} = 0.5$.

Problems. Perform a unit conversion to solve each problem.

① Daniel has a collection of 40 balls that weigh 8 oz each. How many pounds does the whole collection weigh?

② Jessie is running 50 m sprints. How many sprints will she need to run in order to run a total of 0.6 kilometers?

③ A moving company needs to transport 12 desks. Each desk weighs 500 pounds. What is the combined weight of the desks in tons?

④ Alicia pours 2 gallons of punch equally into 8 glasses. How many fluid ounces of punch does each glass contain?

⑤ Marcus is 55 in. tall. How many more inches does Marcus need to grow in order to be 6 feet tall?

9.7 Lines and angles

When two lines **intersect**, they form an **angle**. The point where the lines intersect is called a **vertex**. (Note that the plural form of vertex is **vertices**.) There are three types of angles:

- A **right** angle equals 90°. A right angle forms when **perpendicular** lines intersect. We use the symbol □ to indicate a right angle and the symbol $\perp$ to indicate that two lines are perpendicular.
- An **acute** angle is less than 90°.
- An **obtuse** angle is greater than 90°.

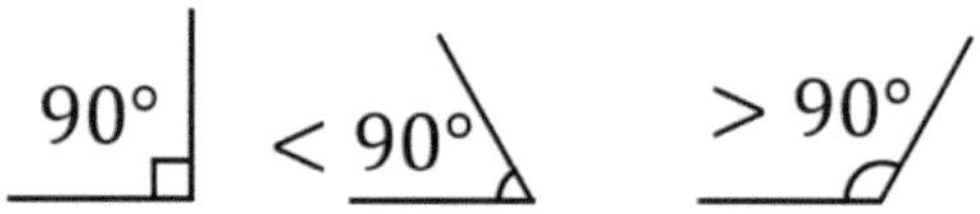

90° $< 90^\circ$ $> 90^\circ$

right acute obtuse

If two angles have the same degree measure, they are said to be **congruent**. Similarly, if two line segments have the same length, they are said to be **congruent**. The symbol $\cong$ indicates congruence. If two or more line segments have one short line through them, this indicates that those line segments are congruent. If there are two pairs of congruent line segments, there will be one short line through one pair and two short lines through the other pair, like the diagram below. If two angles are congruent, this will be shown by drawing a double (or triple) arc for each angle.

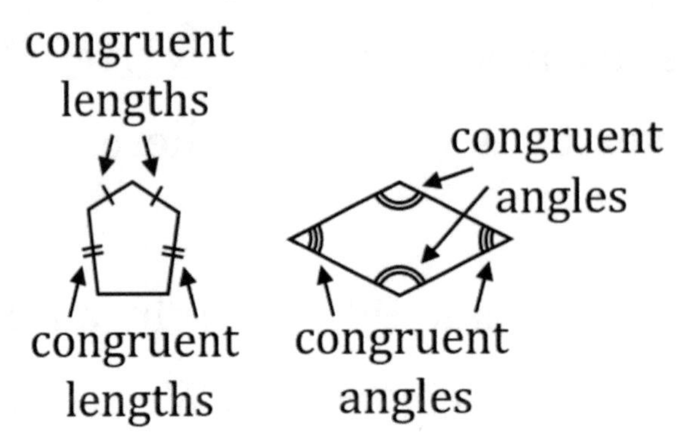

Example. (A) Indicate if each angle is acute, right, or obtuse.

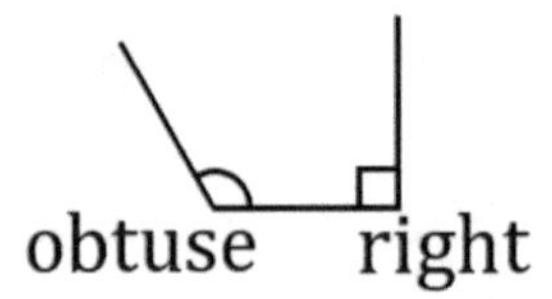

Problems. Indicate if each angle is acute, right, or obtuse.

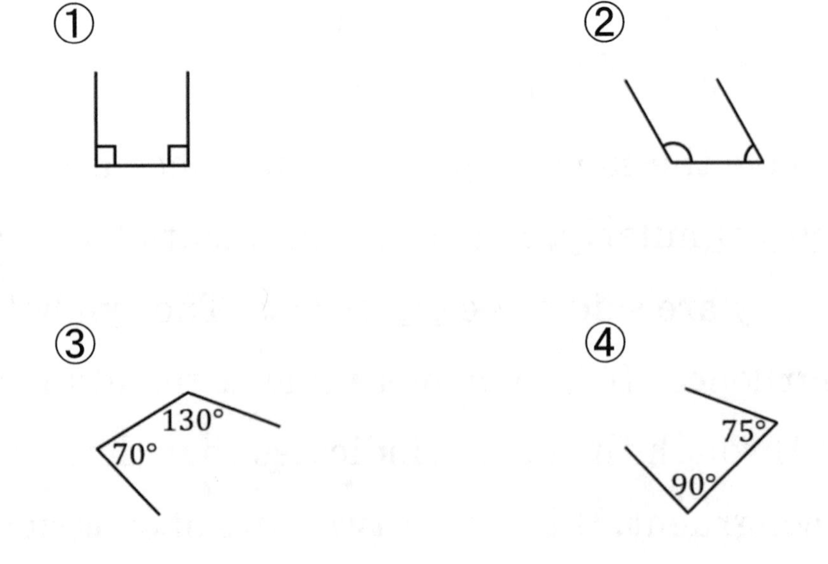

9.8 Polygons

A **polygon** is a closed plane figure bounded by line segments. The name of a polygon is based on the number of sides. A polygon has equal numbers of sides, angles, and vertices. A polygon is **regular** if all of its sides are congruent **and** if all of its angles are congruent. Otherwise, it is **irregular**. A polygon is **equilateral** if the sides are congruent and is **equiangular** if the angles are congruent; it is regular if it is both equilateral and equiangular.

- A **triangle** is a polygon with 3 sides.
- A **quadrilateral** is a polygon with 4 sides.
- A **pentagon** is a polygon with 5 sides.
- A **hexagon** is a polygon with 6 sides.
- A **heptagon** is a polygon with 7 sides.
- An **octagon** is a polygon with 8 sides.
- A **nonagon** is a polygon with 9 sides.
- A **decagon** is a polygon with 10 sides.

Example. (A) Name the shape. Is it regular or irregular?

It is an irregular triangle. It has 3 sides. Only 2 are congruent.

Problems. Name and describe each shape below.

①

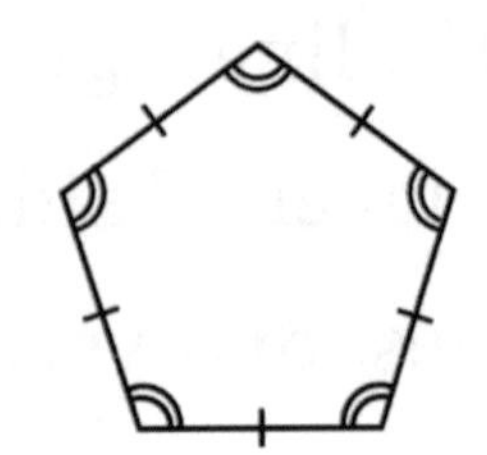

②

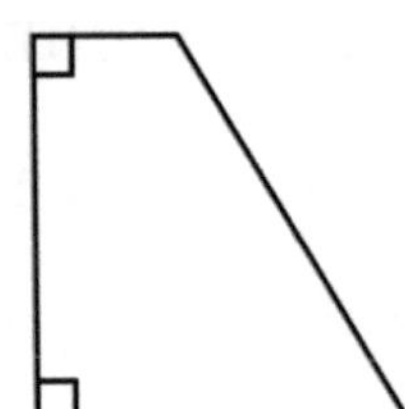

③

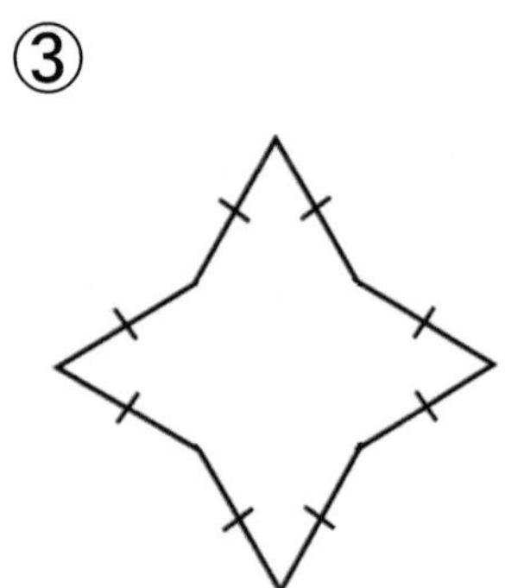

④

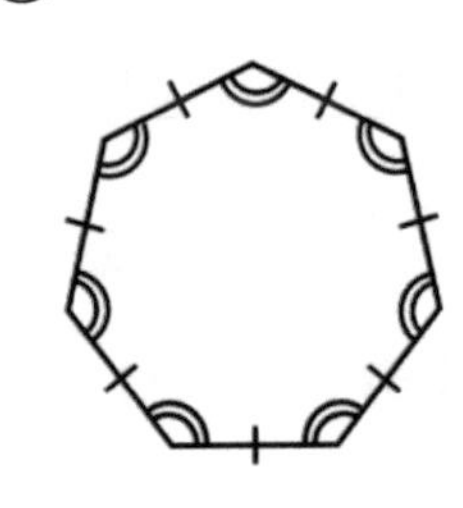

⑤

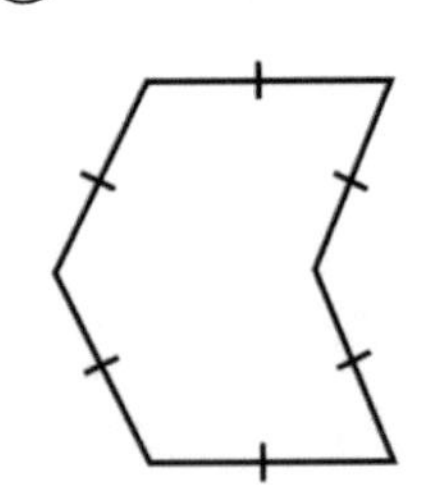

⑥

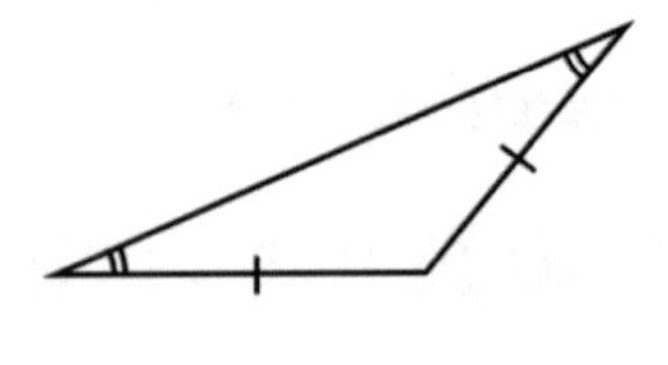

⑦

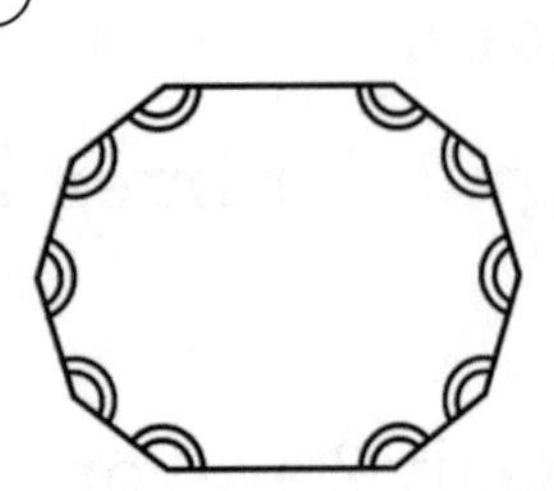

⑧

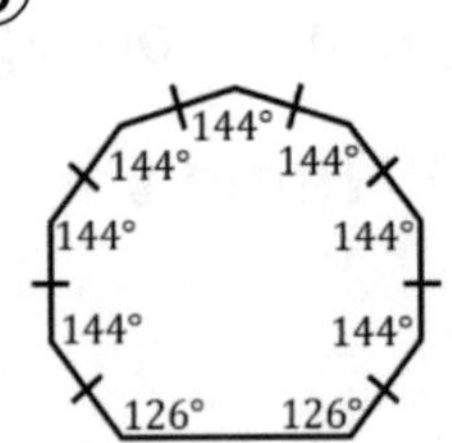

9.9 Triangles

Every triangle is one of the following kinds:

- A **right** triangle has one 90° angle (in addition to two acute angles)
- An **obtuse** triangle has one angle greater than 90° (in addition to two acute angles).
- An **acute** triangle has three angles less than 90°.

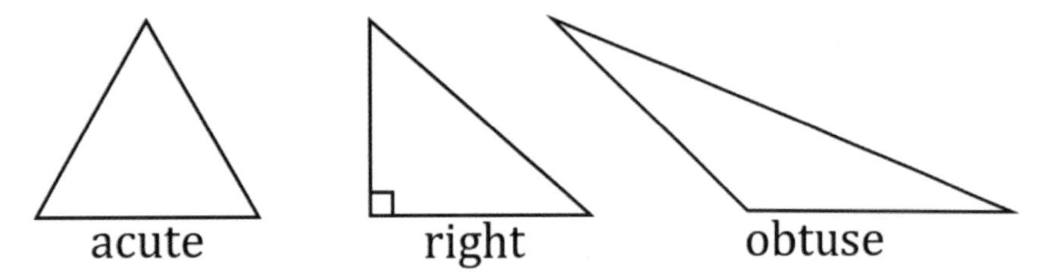

The following terms relate to how many sides (if any) of a triangle are congruent:

- An **isosceles** triangle has at least two congruent sides.
- An **equilateral** triangle has three congruent sides. An equilateral triangle is also equiangular and therefore regular. (However, as we saw in the Sec. 9.8 exercises, a polygon with more than three sides could be equilateral and irregular.) An equilateral triangle is isosceles, but an isosceles triangle isn't necessarily equilateral.
- A **scalene** triangle doesn't have any congruent sides.

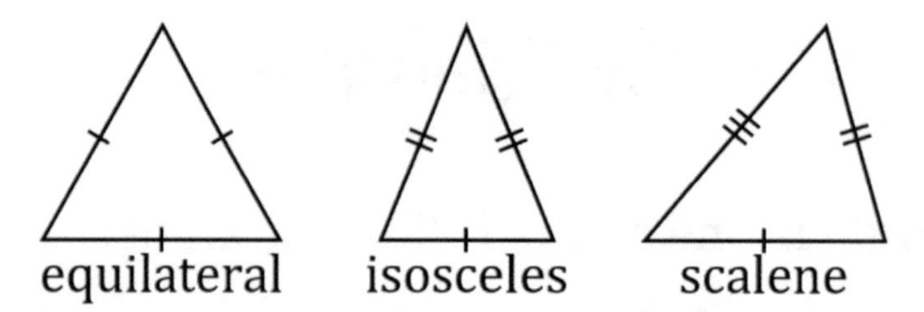

If two angles are congruent, the sides opposite to those angles are also congruent, such that the triangle is isosceles. If all three angles are congruent, the angles are 60° and the triangle is equilateral.

Example. (A) Describe the triangle below.

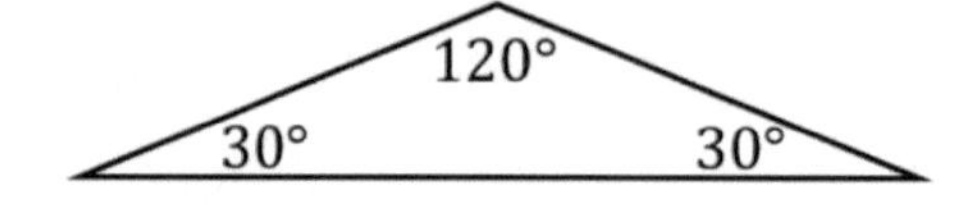

It is an obtuse isosceles triangle.

Problems. For each triangle below, indicate if it is acute, right, obtuse, scalene, isosceles, or equilateral.

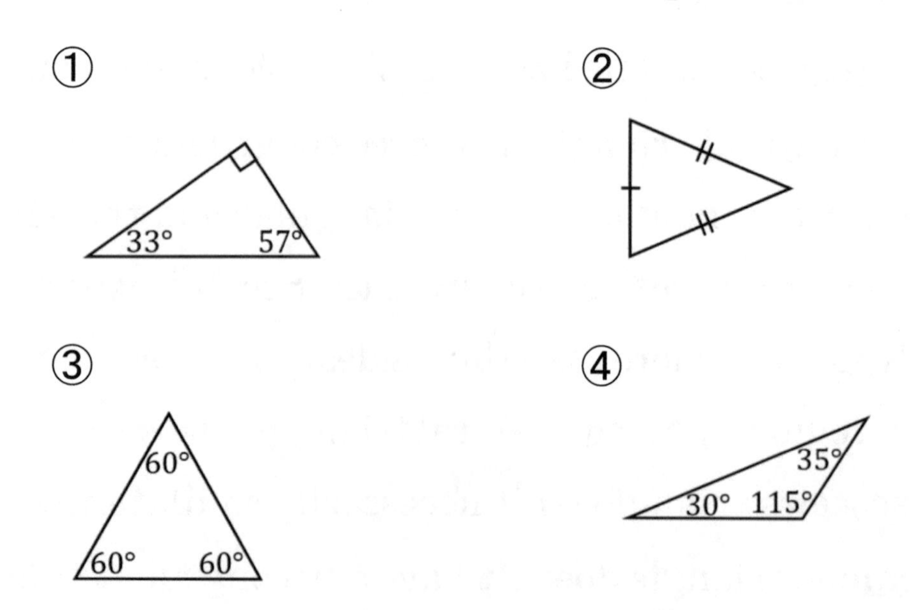

9.10 Quadrilaterals

A **quadrilateral** is a polygon with four sides. Two special quadrilaterals include:

- A **trapezoid** is a quadrilateral with one pair of parallel edges.
- A **parallelogram** is a quadrilateral with two pairs of parallel edges.

Special parallelograms include:

- A **rectangle** has 90° angles. It is equiangular.
- A **rhombus** has congruent edge lengths. It is equilateral.
- A **square** has 90° angles and congruent edge lengths. It is both equiangular and equilateral. It is regular.

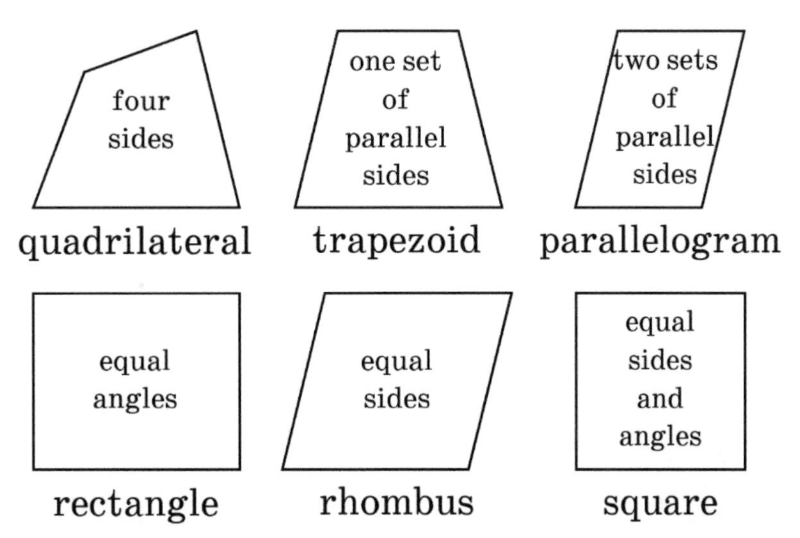

Example. (A) Name the polygon below.

5

3 3

5

It is a parallelogram. It has two pairs of parallel edges.

Problems. Give the most precise name for each polygon.

①

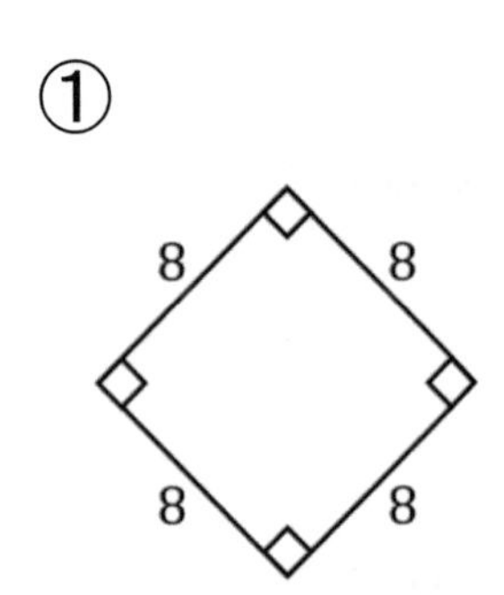

②

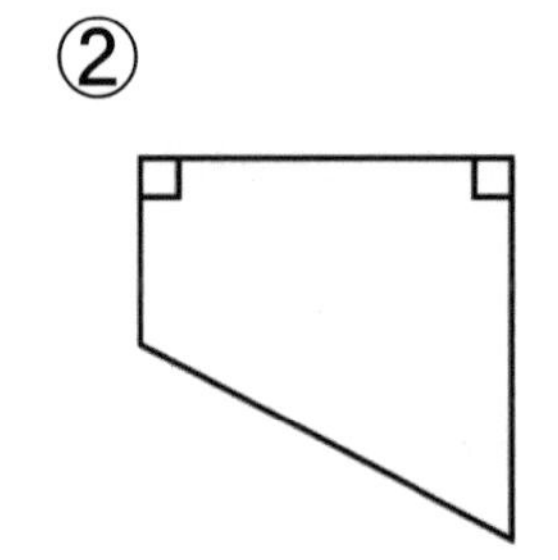

③

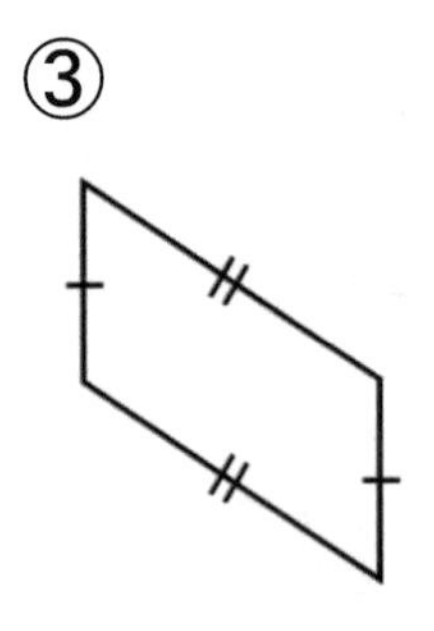

④

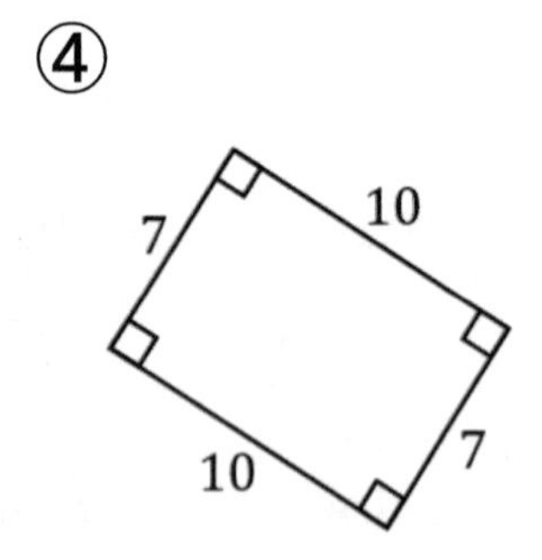

⑤

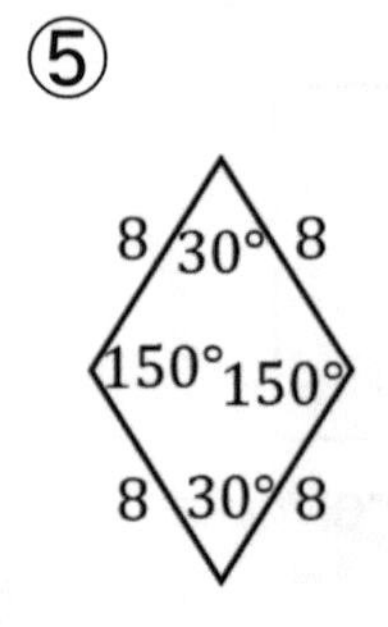

⑥

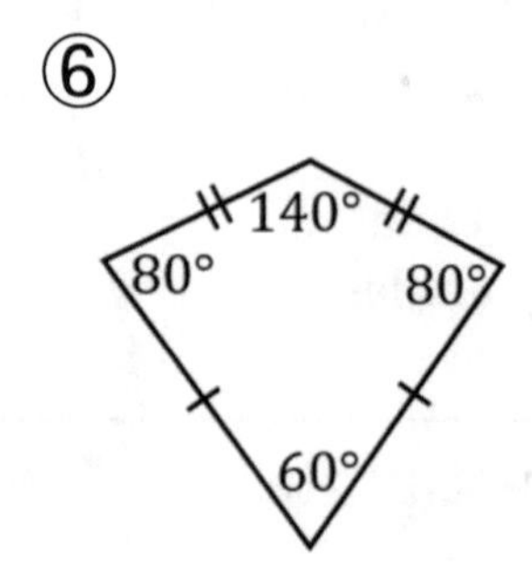

9.11 Perimeter and area

The perimeter of a polygon is the sum of its edge lengths. For example, the perimeter of a rectangle equals twice the length plus twice the width: $P = 2L + 2W$. For other kinds of polygons, add the up the edge lengths.

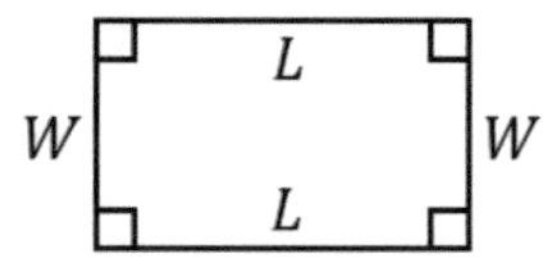

The area of a rectangle equals its length times its width: $A = LW$. For the area of a square, its length equals its width: $A = L^2$. The area of a triangle is one-half its base times its height: $A = \frac{1}{2}bh$.

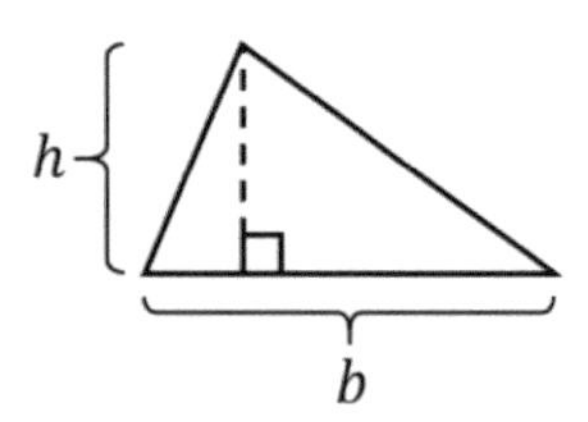

Example. (A) Find the perimeter and area of the rectangle.

Use the formulas for a rectangle with $L = 5$ and $W = 3$:

$$P = 2L + 2W = 2 \times 5 + 2 \times 3 = 10 + 6 = 16$$

$$A = LW = 5 \times 3 = 15$$

Problems. Find the perimeter and area of each polygon.

①

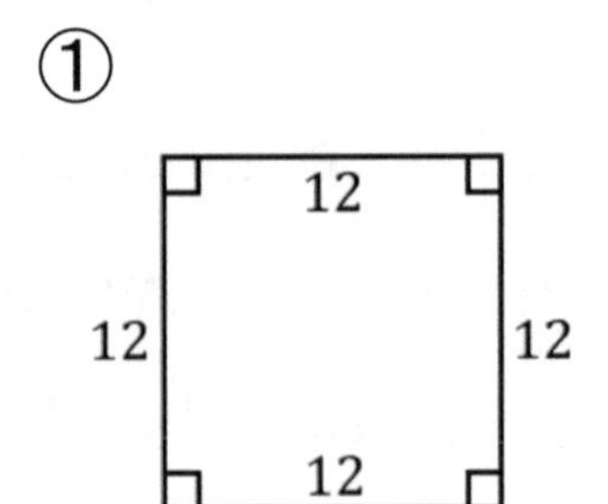

②

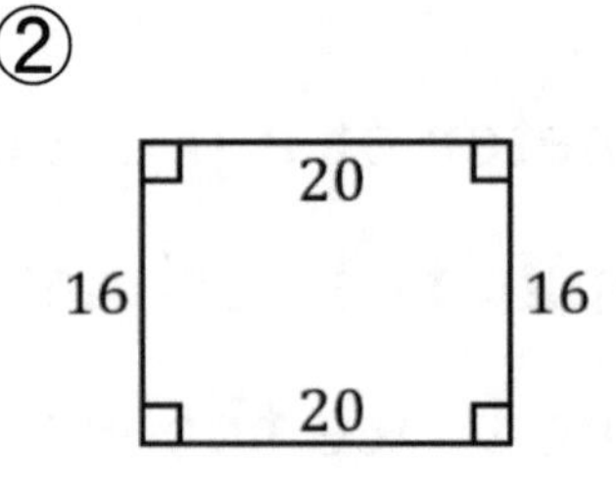

③

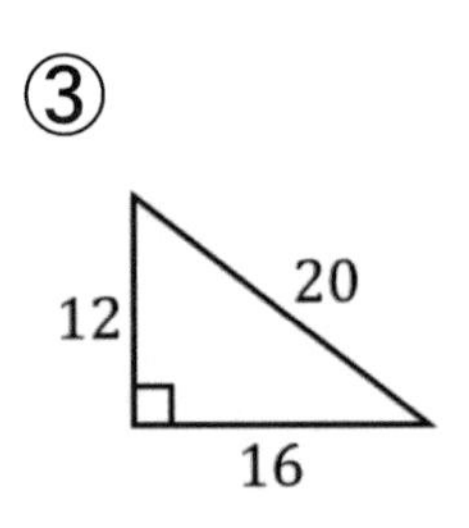

④

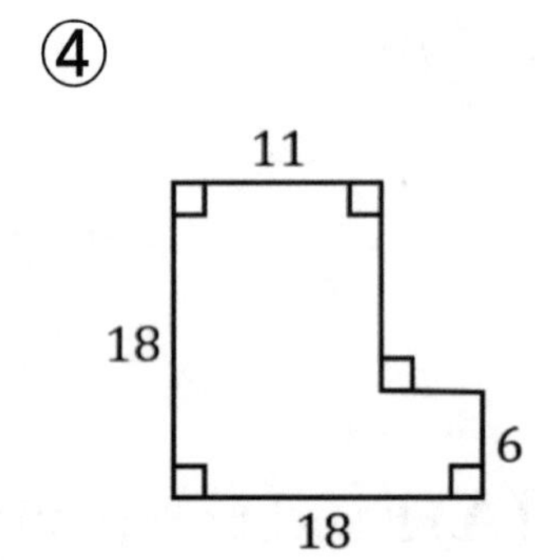

9.12 Volume

A **rectangular prism** (also called a cuboid) has the shape of a rectangular box. It has 6 sides. Each side is a rectangle. The volume of a rectangular prism can be found two ways:

- Use the formula $V = LWH$ (length times width times height).
- First find the area of the base ($A = LW$). Then multiply the base area by the height: $V = AH$. (Note: Some texts and teachers use B for the base area instead of A. By using A for the base area, we hope to avoid confusion with the base of a triangle, b.)

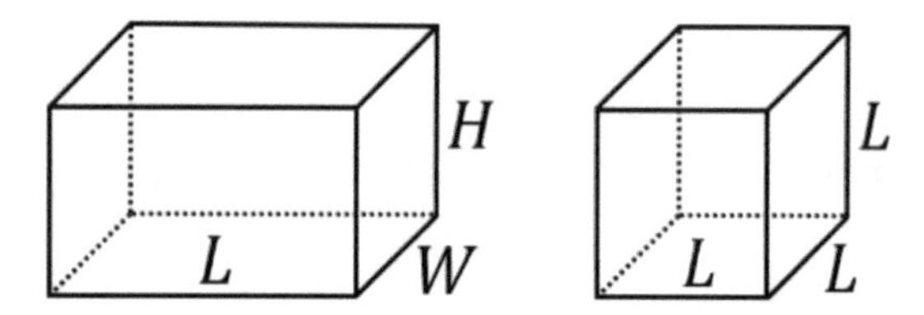

A **cube** is a special case of a rectangular prism where the length, width, and height are equal. The volume of a cube is $V = L^3$ (length times length times length).

Example. (A) Find the volume of a rectangular prism that has a length of 8, a width of 5, and a height of 4.

Use the formula: $V = LWH = 8 \times 5 \times 4 = 40 \times 4 = 160$.

Problems. Find the volume of each shape.

①

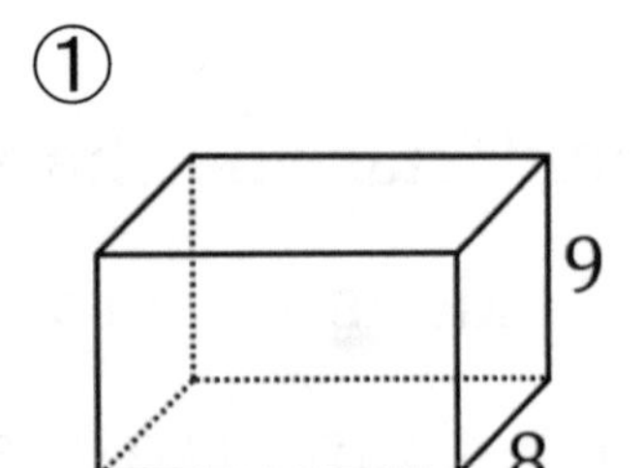

②

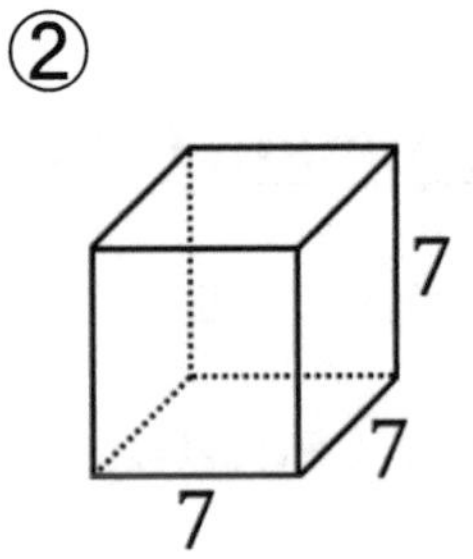

③ A rectangular prism has a length of 12, a width of 10, and a height of 4. What is its volume?

④ The length of a cube is 8. What is its volume?

⑤ A rectangular prism has a square base with an edge length of 6 and a height of 5. What is its volume?

9.13 Visualizing volume

One way to visualize the volume of a rectangular prism is to divide the rectangular prism up into unit cubes. Each unit cube has a volume equal to one cubic unit. Add up the total number of unit cubes to determine the volume of the rectangular prism.

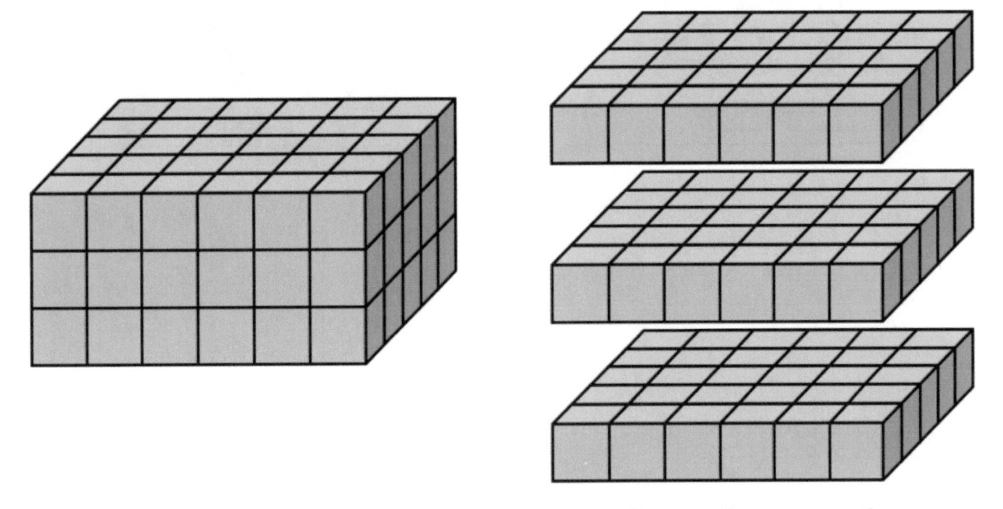

For example, the rectangular prism above has 6 cubes across, 5 cubes deep, and 3 cubs high. There are $6 \times 5 = 30$ cubes in the base and a total of $30 \times 3 = 90$ cubes because each of the three layers has 30 cubes (as shown above on the right). Note that this agrees with the formulas from Sec. 9.12:

$$V = LWH = 6 \times 5 \times 3 = 30 \times 3 = 90$$

or

$$A = LW = 6 \times 5 = 30 \quad \text{and} \quad V = AH = 30 \times 3 = 90$$

Example. (A) Find the volume of the shape below.

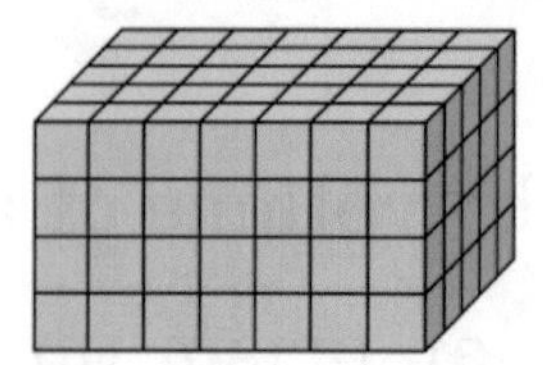

$$V = LWH = 7 \times 5 \times 4 = 35 \times 4 = 140$$

Problems. Find the volume of each rectangular prism.

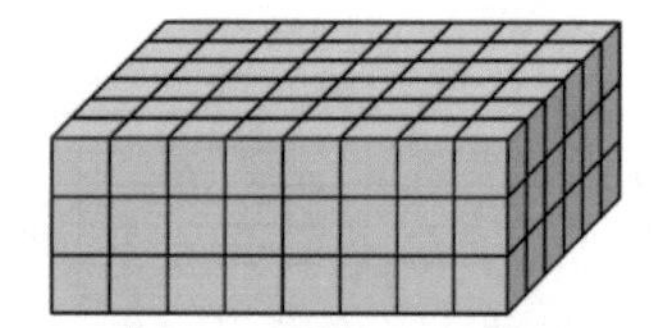

②

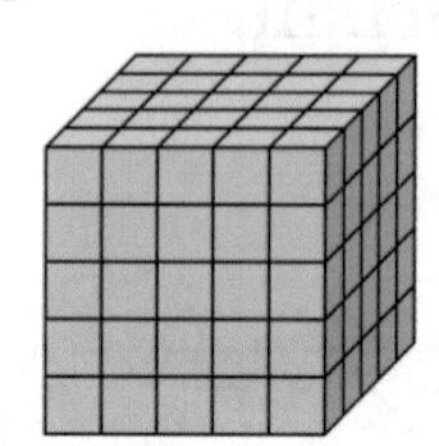

③

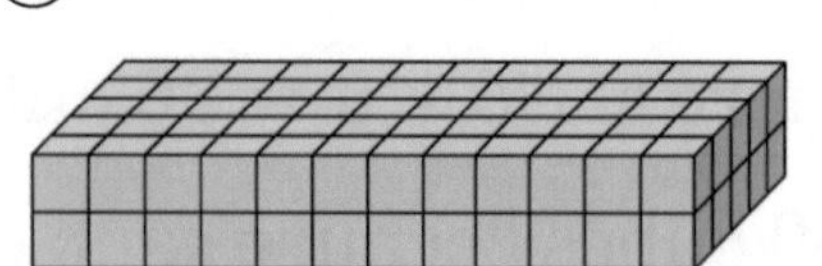

④

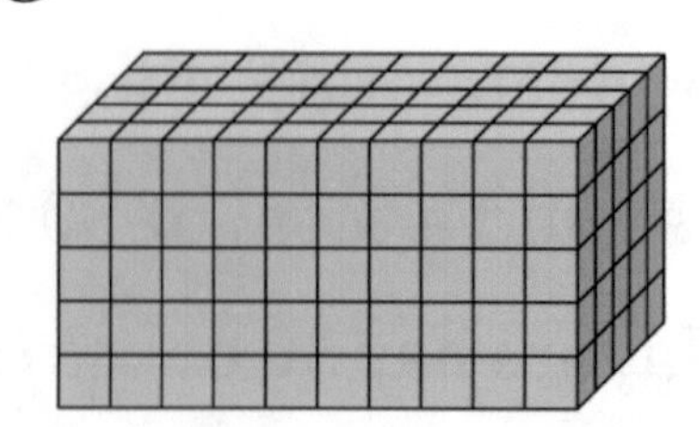

Multiple Choice Questions

① Which device is best to measure the length of a jogging track that surrounds a football field?

(A) odometer (B) ruler (C) tape measure (D) trundle wheel (E) yardstick

② Which device is best to measure the mass of a key?

(A) balance (B) measuring cup (C) ruler (D) truck scale (E) trundle wheel

③ Which could be the length of a tennis court?

(A) 0.15 mi (B) 8 ft (C) 26 yd (D) 78 m (E) 90 in.

④ Which could be the capacity of a coffee mug?

(A) 2 gal (B) 6 qt (C) 8 pt (D) 10 c (E) 12 fl oz

⑤ Which is equivalent to 3 cL?

(A) 0.003 L (B) 0.03 L (C) 0.3 L (D) 30 L (E) 300 L

⑥ Convert 24 ft to yards.

(A) 2 yd (B) 8 yd (C) 48 yd (D) 72 yd (E) 288 yd

⑦ Convert 48 qt to cups.

(A) 6 c (B) 12 c (C) 24 c (D) 96 c (E) 192 c

⑧ Convert 272 oz to pounds.

(A) 17 lb (B) 34 lb (C) 68 lb (D) 2176 lb (E) 4352 lb

⑨ Convert 37.4 cm to meters.

(A) 0.0374 m (B) 0.374 m (C) 3.74 m (D) 374 m (E) 3740 m

⑩ Convert 6.2 kg to g.

(A) 0.0062 g (B) 0.062 g (C) 0.62 g (D) 620 g (E) 6200 g

⑪ Which measurement below is the largest? (Note that a meter is just a few inches longer than a yard.)

(A) 2 m (B) 5 yd (C) 14 ft (D) 87 cm (E) 175 in.

⑫ Which inequality is **<u>NOT</u>** correct?

(A) 3 c > 20 fl oz (B) 2 gal < 9 qt (C) 5 gal > 35 pt

(D) 7 pt < 15 c (E) 9 qt < 30 c

⑬ Melinda has 3 chains that are 2 yd long and 4 chains that are 18 in. long. What is the total length of all of the chains?

(A) 24 ft (B) 78 in. (C) 78 yd (D) 288 ft (E) 288 yd

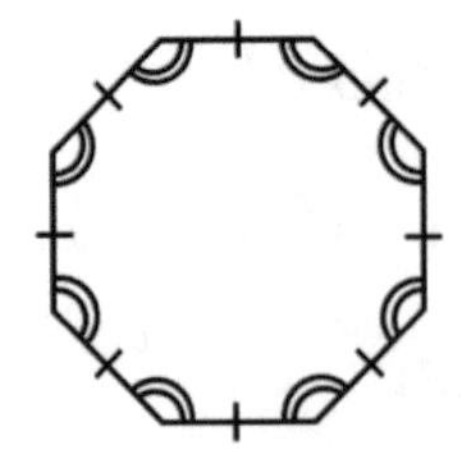

⑭ Which shape appears above?

(A) decagon (B) hexagon (C) nonagon (D) octagon

⑮ Which best describes the shape above?

(A) equiangular only (B) equilateral only (C) regular

(D) none of the above terms describe this shape

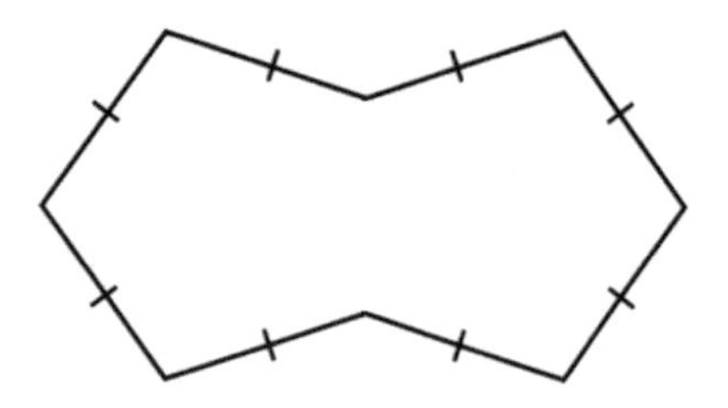

⑯ Which shape appears above?

(A) heptagon (B) nonagon (C) octagon (D) pentagon

⑰ Which best describes the shape above?

(A) equiangular only (B) equilateral only (C) regular

(D) none of the above terms describe this shape

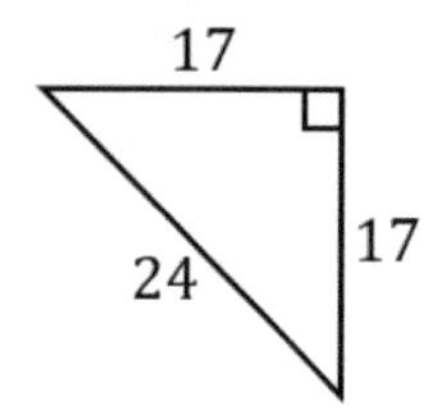

⑱ Which best describes the triangle above?

(A) acute, scalene (B) right, isosceles (C) right, equilateral

(D) right, scalene (E) obtuse, isosceles

⑲ What is the **perimeter** of the triangle above?

(A) 41 (B) 58 (C) 65 (D) 289 (E) 408

⑳ What is the **area** of the triangle above?

(A) 34 (B) 144.5 (C) 204 (D) 289 (E) 408

㉑ What is the **area** of a triangle with a base of 9 and a height of 6?

(A) 7.5 (B) 15 (C) 27 (D) 30 (E) 54

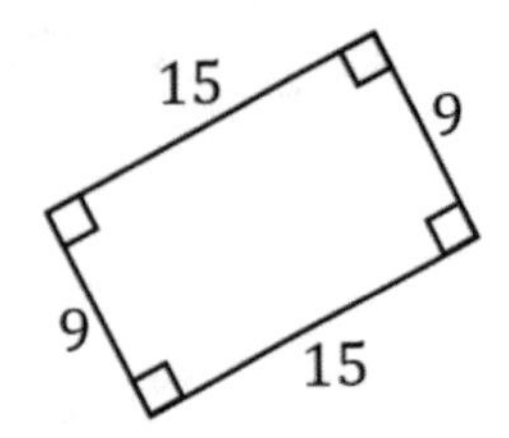

㉒ Which best describes the shape above?

(A) pentagon (B) rectangle (C) rhombus

(D) square (E) trapezoid

㉓ What is the **perimeter** of the shape above?

(A) 24 (B) 33 (C) 39 (D) 48 (E) 81

㉔ What is the **area** of the shape above?

(A) 67.5 (B) 112.5 (C) 81 (D) 135 (E) 225

10
7
8
3

㉕ What is the **area** of the shape above?

(A) 21 (B) 56 (C) 59 (D) 80 (E) 94

㉖ The edge of a cube is 4. What is its **volume**?

(A) 8 (B) 12 (C) 16 (D) 32 (E) 64

㉗ What is the **area** of a rectangle with length 7 and width 4?

(A) 11 (B) 14 (C) 22 (D) 28 (E) 32

㉘ What is the **volume** of a rectangular prism with length 7, width 6, and height 5?

(A) 36 (B) 42 (C) 72 (D) 105 (E) 210

-10-

FINANCE

10.1 Coins

The values of the most common coins in the United States are tabulated below in terms of dollars ($). Note that 100 cents (¢) is equivalent to one dollar ($): 100¢ = $1.

Penny	Nickel	Dime	Quarter
$0.01	$0.05	$0.10	$0.25

It may help to review Chapter 4 (arithmetic with decimals).

Example. (A) What is the total value of 7 quarters and 4 dimes in terms of dollars?

$$\begin{array}{r} {}^{3} \\ 0.25 \\ \times 7 \\ \hline 1.75 \end{array} \qquad \begin{array}{r} 0.10 \\ \times 4 \\ \hline 0.40 \end{array} \qquad \begin{array}{r} {}^{1} \\ 1.75 \\ +0.40 \\ \hline 2.15 \end{array}$$

$$7 \times \$0.25 + 4 \times \$0.10 = \$1.75 + \$0.40 = \$2.15$$

Problems. Determine the total value of the coins in dollars.

① 37 quarters

② 63 nickels

③ 28 nickels and 34 pennies

④ 17 dimes and 35 nickels

⑤ 14 quarters and 9 dimes

⑥ 73 quarters, 42 dimes, and 29 nickels

⑦ 22 quarters, 11 dimes, 17 nickels, and 56 pennies

⑧ 45 quarters, 24 dimes, 18 nickels, and 78 pennies

10.2 Tax on sales and property

Sales tax and property tax are two common kinds of taxes that state or local governments collect that relate to types of purchases that people make:

- **Sales tax** is money that is paid when a customer pays for goods or services. For example, when a customer makes a purchase in a store, the customer is charged sales tax in addition to the cost of the items.
- **Property tax** is money that is paid to the government for owning property. For example, most homeowners pay a property tax once each year.

Examples. (A) A customer buys a dress that costs \$36. The customer pays a total amount of \$38.88. How much tax did the customer pay? Which kind of tax is this?

Subtract \$36.00 from \$38.88.

$$\begin{array}{r} 38.88 \\ -\ 36.00 \\ \hline 2.88 \end{array}$$

The tax is \$2.88. This is a sales tax.

(B) A house has a value of $150,000. The county charges an annual tax of $20 for every $1000 of the home's value. How much is the annual tax for this house? Which kind of tax is this?

There are 150 thousands in $150,000. The county charges $20 for each thousand dollars. Multiply 150 by $20. Since $15 \times 2 = 30$ and since 150×20 has two zeroes, the answer is 150 × $20 = $3000 (thirty plus two zeroes). The annual tax is $3000. This is a property tax.

Problems. Solve each word problem.

① A customer buys a book that costs $14. The customer pays a total amount of $15.19. How much tax did the customer pay? Which kind of tax is this? Will the customer continue to pay this tax every year?

② A person purchases land for $18,000. The person has to pay a total of $18,500. How much tax did the customer pay? Which kind of tax is this? Will the customer continue to pay this tax every year?

③ A customer buys a pen that costs $3. The customer is charged a tax of $0.08 for each $1 spent. How much is the tax? Which kind of tax is this? What is the total amount that the customer must pay?

④ A family owns a house with a value of $80,000. The city charges an annual tax of $25 for every $1000 of the home's value. How much is the annual tax for this house? Which kind of tax is this?

⑤ A customer buys clothing that costs $150. The customer is charged a tax of $1 for each $10 spent. How much is the total tax? Which kind of tax is this? What is the total amount that the customer must pay?

10.3 Tax on earnings

Income tax and payroll tax are two common kinds of taxes that are paid based on money that is earned:

- **Income tax** is money that is paid to the government based on earnings. Each year, federal income tax must be filed and paid to the US government. Some state and local governments also collect income tax.
- **Payroll tax** is money that an employer withholds from an employee's paycheck to pay the government.

Examples. **(A)** Walter earns \$600 per week. His employer withholds \$7 in taxes for every \$100 earned. How much does Walter's employer withhold in taxes each week? Which kind of tax is this?

There are 6 hundreds in \$600. The employer withholds \$7 for each hundred dollars earned. The total amount withheld each week is $6 \times \$7 = \$42$. This is a payroll tax.

(B) Susan earned \$36,000 last year. She filed her taxes with the federal government and paid \$5 in taxes for every \$100 earned. What total amount did Susan pay in taxes? Which kind of tax is this?

There are 360 hundreds in \$36,000 (because $360 \times \$100 = \$36{,}000$). Susan paid \$5 for each hundred dollars earned. The total tax is $360 \times \$5 = \$1800$. This is a federal income tax.

Problems. Solve each word problem.

① Janet earns \$1800 per month. Her employer withholds \$6 in taxes for every \$100 earned. How much does Janet's employer withhold in taxes each month? Which kind of tax is this?

② Wesley earned \$42,000 last year. When Wesley filed his taxes, he paid \$7 in federal taxes for every \$100 earned and \$2 in state taxes for every \$100 earned. What total amount did Wesley pay in taxes? Which kind of tax is this?

10.4 Gross and net income

The **gross** income is the total amount earned. The **net** income is the amount earned after taxes (and any other deductions). For example, if a person earns \$800 per week before paying taxes of \$70 per week, the gross income is \$800 per week and the net income is $\$800 - \$70 = \$730$ per week.

Example. (A) Molly's gross earnings are \$2100 per month. Molly's employer withholds \$150 for federal income taxes and \$50 for state income taxes. What is Molly's net income?

Subtract the payroll taxes from the gross income:

$$\$2100 - \$150 - \$50 = \$1900$$

(Note: These are payroll taxes because they were withheld by the employer and the employer paid this money to the government. This money was withheld by the employer on behalf of the employee for federal and state income taxes that the employee must pay.)

Problems. Solve each word problem.

① Tam's gross income is \$360 per week. Tam has \$75 per week withheld in taxes. What is Tam's net income?

② Jim's net income is $1540 per month. Each month, Jim has $135 withheld for federal income taxes and $28 withheld for state income taxes. What is Jim's gross income?

③ Tamara earned $640 for a part-time job. Her employer withheld $48 for federal income tax. There was also state income tax withheld. Tamara's net income was $578. How much was withheld for state income taxes?

④ Maurice earned $1250 last month. His employer withheld $83 for federal income tax, $26 for state income tax, and $9 for local income tax. What is Maurice's net income?

10.5 Budgets

A **budget** is a list of expected income and expenses for a period of time. A budget is said to be **balanced** if the total income equals the total expenses.

Example. (A) Tori's income and expenses for next week are listed below. Is Tori's budget balanced? If not, what might Tori do to balance her budget?

Income	
Chores	\$25
Babysitting	\$60

Expenses	
Dining	\$20
Shopping	\$35
Entertainment	\$15
Savings	\$10

- Tori's total income is $\$25 + \$60 = \$85$.
- Tori's total expenses are $\$20 + \$35 + \$15 + \$10 = \$80$.
- Tori's budget isn't balanced because \$85 is different from \$80.
- Tori could add \$5 more to her savings. Tori could save \$15 per week instead of \$10. This would bring Tori's total expenses up to $\$80 + \$5 = \$85$.

Problems. Answer the questions about each budget.

① Carl's income and expenses are shown below, except for the amount of savings. How much should Carl save?

Income	
Allowance	$40
Dog walking	$32

Expenses	
Transportation	$12
Supplies	$24
Fees	$17
Savings	

② Claire's monthly income and expenses are shown below, except for her income for tutoring. How much does Claire need to earn for tutoring each month? How long will it take for Claire to save $150?

Income	
Chores	$50
Yard work	$25
Tutoring	

Expenses	
Books	$18
Ink/paper	$22
Software	$55
Savings	$25

10.6 Debits and credits

When keeping written records of account activity:

- A <u>credit</u> is money that goes into an account.
- A <u>debit</u> is money that comes out of an account.

Example. (A) If you deposit your birthday money into your savings account, is this a credit or debit?

This is a credit because it increases your account balance. You put this money into your savings account.

(B) If you withdraw $45 from your checking account, is this a credit or debit?

This is a debit because it decreases your account balance. You took this money out of your checking account.

Problems. Is the action a debit or credit to your account?

① You withdraw $40 from an ATM.

② Your paycheck is direct deposited.

③ You buy a souvenir using your debit card.

④ You write a check for $15 to pay a bill.

10.7 Keeping track of money

It is wise to keep a written record of finances. Even if you use a bank with a checking or savings account, although the bank will give you written or online statements of your account activity, it is wise to maintain your own written records for comparison. A good record will include:

- dates showing when each transaction is made
- a brief yet informative description of each transaction
- a way of clearly showing whether a transaction is a debit or credit (see the sample record that follows)
- an update of the current balance that results from each transaction

Date	Description	Credit	Debit	Balance
	Beginning balance			\$63
4/3	Allowance	\$12		\$75
4/4	Snack		\$3	\$72
4/7	Birthday money	\$50		\$122
4/8	Shopping		\$30	\$92

Example. (A) If \$15 is spent on movies, how would this be entered on the record above?

It is a debit. The new balance would be $\$92 - \$15 = \$77$.

Problem. ① Fill in the rightmost column (called Balance) for the record below.

Date	Description	Credit	Debit	Balance
	Beginning balance			$231
9/2	Bus fare		$3	
9/3	Allowance	$25		
9/5	Movie		$9	
9/5	Bookstore		$16	
9/7	Tutoring	$35		
9/8	Club fees		$4	
9/10	Allowance	$25		
9/10	Supplies		$23	
9/12	Miscellaneous		$18	

Multiple Choice Questions

① What is value of 41 quarters?

(A) \$2.05 (B) \$8.25 (C) \$9.25 (D) \$10.25 (E) \$12.50

② What is value of 67 nickels?

(A) \$3.35 (B) \$3.45 (C) \$3.55 (D) \$6.35 (E) \$6.70

③ How many quarters make \$8.75?

(A) 33 (B) 35 (C) 37 (D) 330 (E) 350

④ What is the total value of 11 quarters, 3 dimes, 7 nickels, and 18 pennies?

(A) \$2.58 (B) \$3.53 (C) \$3.58 (D) \$3.77 (E) \$3.90

⑤ Which tax does a customer pay when a customer buys a pair of shoes?

(A) income tax (B) payroll tax (C) property tax (D) sales tax

⑥ Which tax requires a person to file taxes with the federal government at the end of the year based on earnings?

(A) income tax (B) payroll tax (C) property tax (D) sales tax

⑦ Which tax does a homeowner pay annually based on the value of the home?

(A) income tax (B) payroll tax (C) property tax (D) sales tax

⑧ A customer buys a pair of earrings that cost $29.50. The total amount paid is $31.86. What is the sales tax?

(A) $1.36 (B) $1.86 (C) $2.36 (D) $2.86 (E) $61.36

⑨ A customer buys $7 worth of fruit. The store charges $0.05 of sales tax for each $1 spent. What is the sales tax?

(A) $0.005 (B) $0.05 (C) $0.25 (D) $0.35 (E) $0.70

⑩ A county charges $45 of property tax for each $1000 of the value of a home. What is the property tax for a home with a value of $90,000?

(A) $365 (B) $405 (C) $3650 (D) $4050 (E) $4500

⑪ Dee earned $1400. Her employer withheld $79 for federal income tax and $23 for state income tax. What is Dee's net income?

(A) $1298 (B) $1344 (C) $1398 (D) $1456 (E) $1502

⑫ Which of the following is a **credit** to your bank account?

(A) ATM withdrawal (B) debit card purchase

(C) payroll deposit (D) writing a check

⑬ Which of the following is a **debit** from your bank account?

(A) ATM withdrawal (B) deposit funds in person

(C) deposit a roll of quarters (D) payroll deposit

Income	
Allowance	\$40
Baby sitting	\$35
Pet sitting	\$15
Tutoring	\$25

Expenses	
Transportation	\$14
Shopping	\$53
Entertainment	\$22
Savings	\$16

⑭ Which statement is correct about the budget above?

(A) The budget is balanced. (B) Income exceeds expenses.

(C) Expenses exceed income. (D) There is no income.

⑮ What could be done to balance the budget above?

(A) Nothing. It is balanced. (B) Add \$10 to savings.

(C) Subtract \$10 from savings. (D) Earn \$10 more pet sitting.

⑯ If Elias saves \$16 every month, how long will it take Elias to save \$80?

(A) 5 months (B) 6 months (C) 64 months (D) 96 months

⑰ Sadie's bank account balance is currently \$327. Sadie uses her debit card to buy a gift for \$58. What is Sadie's bank account balance after making this transaction?

(A) \$269 (B) \$271 (C) \$279 (D) \$281 (E) \$385

ANSWER KEY

Chapter 1: Multiplication

Page 6

① 24	② 35	③ 72
④ 49	⑤ 54	⑥ 36
⑦ 40	⑧ 28	⑨ 30
⑩ 63	⑪ 64	⑫ 35
⑬ 36	⑭ 36	⑮ 48
⑯ 45	⑰ 42	⑱ 20
⑲ 32	⑳ 25	㉑ 81
㉒ 42	㉓ 72	㉔ 56
㉕ 16	㉖ 63	㉗ 27
㉘ 48	㉙ 24	㉚ 54

Page 9

① $\begin{array}{r} {}^{2}\\ 47 \\ \times 3 \\ \hline 141 \end{array}$

② $\begin{array}{r} {}^{4}\\ 29 \\ \times 5 \\ \hline 145 \end{array}$

③ $\begin{array}{r} {}^{2}\\ 54 \\ \times 6 \\ \hline 324 \end{array}$

④ $\begin{array}{r} 12 \\ \times 4 \\ \hline 48 \end{array}$

⑤ $\begin{array}{r} 31 \\ \times 8 \\ \hline 248 \end{array}$

⑥ $\begin{array}{r} {}^{3}\\ 75 \\ \times 7 \\ \hline 525 \end{array}$

⑦
$$\begin{array}{r} {}^{7}\\ 68 \\ \times 9 \\ \hline 612 \end{array}$$

⑧
$$\begin{array}{r} 83 \\ \times 2 \\ \hline 166 \end{array}$$

⑨
$$\begin{array}{r} {}^{4}\\ 46 \\ \times 7 \\ \hline 322 \end{array}$$

⑩
$$\begin{array}{r} {}^{5}\\ 97 \\ \times 8 \\ \hline 776 \end{array}$$

Page 11

①
$$\begin{array}{r} {}^{3\,1}\\ 263 \\ \times 5 \\ \hline 1315 \end{array}$$

②
$$\begin{array}{r} {}^{1\,1}\\ 459 \\ \times 2 \\ \hline 918 \end{array}$$

③
$$\begin{array}{r} {}^{1\,4}\\ 617 \\ \times 7 \\ \hline 4319 \end{array}$$

④
$$\begin{array}{r} {}^{6\,3}\\ 384 \\ \times 8 \\ \hline 3072 \end{array}$$

⑤
$$\begin{array}{r} {}^{1\,4}\\ 928 \\ \times 6 \\ \hline 5568 \end{array}$$

⑥
$$\begin{array}{r} {}^{2\,1}\\ 875 \\ \times 3 \\ \hline 2625 \end{array}$$

⑦
$$\begin{array}{r} {}^{3}\\ 792 \\ \times 4 \\ \hline 3168 \end{array}$$

⑧
$$\begin{array}{r} {}^{3\,5}\\ 536 \\ \times 9 \\ \hline 4824 \end{array}$$

Page 14

①
$$\begin{array}{r} {}^{1}\\ 53 \\ \times 24 \\ \hline 212 \\ 1060 \\ \hline 1272 \end{array}$$

②
$$\begin{array}{r} {}^{4}\\ 37 \\ \times 16 \\ \hline 222 \\ 370 \\ \hline 592 \end{array}$$

③
3
4
68
× 45
340
2720
3060

④
1
92
× 38
736
2760
3496

⑤
4
5
36
× 79
324
2520
2844

⑥
81
× 50
0
4050
4050

Page 15

⑦
4
88
× 51
88
4400
4488

⑧
2
2
54
× 65
270
3240
3510

⑨
1
1
46
× 33
138
1380
1518

⑩
1
69
× 12
138
690
828

⑪
1
4
95
× 28
760
1900
2660

⑫
2
2
77
× 44
308
3080
3388

Page 17

①
4 3
3 2
386
× 54
1,544
19,300
20,844

②
1 1
4 5
158
× 27
1,106
3,160
4,266

Page 18

③
$$\begin{array}{r} \scriptstyle 2\,7 \\ \scriptstyle 2\; \\ 629 \\ \times\ 83 \\ \hline 1{,}887 \\ 50{,}320 \\ \hline 52{,}207 \end{array}$$

④
$$\begin{array}{r} \scriptstyle 4\,1\; \\ 473 \\ \times\ 16 \\ \hline 2{,}838 \\ 4{,}730 \\ \hline 7{,}568 \end{array}$$

⑤
$$\begin{array}{r} \scriptstyle 2\;\; \\ \scriptstyle 2\;\; \\ 841 \\ \times\ 75 \\ \hline 4{,}205 \\ 58{,}870 \\ \hline 63{,}075 \end{array}$$

⑥
$$\begin{array}{r} \scriptstyle 3\,2\; \\ \scriptstyle 1\,1\; \\ 585 \\ \times\ 42 \\ \hline 1{,}170 \\ 23{,}400 \\ \hline 24{,}570 \end{array}$$

⑦
$$\begin{array}{r} \scriptstyle 2\;\; \\ \scriptstyle 1\;\; \\ 730 \\ \times\ 96 \\ \hline 4{,}380 \\ 65{,}700 \\ \hline 70{,}080 \end{array}$$

⑧
$$\begin{array}{r} 612 \\ \times\ 30 \\ \hline 0 \\ 18{,}360 \\ \hline 18{,}360 \end{array}$$

Page 19

⑨
$$\begin{array}{r} \scriptstyle 3\,2\;\; \\ 264 \\ \times\ 61 \\ \hline 264 \\ 15{,}840 \\ \hline 16{,}104 \end{array}$$

⑩
$$\begin{array}{r} \scriptstyle 3\; \\ \scriptstyle 5\; \\ 407 \\ \times\ 58 \\ \hline 3{,}256 \\ 20{,}350 \\ \hline 23{,}606 \end{array}$$

⑪
$$\begin{array}{r} \scriptstyle 3\,1\; \\ \scriptstyle 3\,1\; \\ 593 \\ \times\ 44 \\ \hline 2{,}372 \\ 23{,}720 \\ \hline 26{,}092 \end{array}$$

⑫
$$\begin{array}{r} \scriptstyle 6\,4\; \\ \scriptstyle 7\,5\; \\ 986 \\ \times\ 79 \\ \hline 8{,}874 \\ 69{,}020 \\ \hline 77{,}894 \end{array}$$

⑬
$$\begin{array}{r} \scriptstyle 1\,2\; \\ \scriptstyle 3\,4\; \\ 757 \\ \times\ 37 \\ \hline 5{,}299 \\ 22{,}710 \\ \hline 28{,}009 \end{array}$$

⑭
$$\begin{array}{r} \scriptstyle 4\,1\; \\ \scriptstyle 3\,1\; \\ 852 \\ \times\ 86 \\ \hline 5{,}112 \\ 68{,}160 \\ \hline 73{,}272 \end{array}$$

Page 21

① $$\begin{array}{r} {\scriptstyle 5\,2} \\ 584 \\ \times\, 6 \\ \hline \boxed{3}504 \end{array}$$

② $$\begin{array}{r} {\scriptstyle 7\,4} \\ \boxed{2}95 \\ \times\, 8 \\ \hline 2360 \end{array}$$

③ $$\begin{array}{r} {\scriptstyle 1\,2} \\ 3\boxed{4}7 \\ \times\, 4 \\ \hline 1388 \end{array}$$

④ $$\begin{array}{r} {\scriptstyle 2\,1} \\ 942 \\ \times\, \boxed{7} \\ \hline 6594 \end{array}$$

⑤ $$\begin{array}{r} {\scriptstyle 2\,4} \\ {\scriptstyle 1\,2} \\ \boxed{2}59 \\ \times\, 53 \\ \hline \boxed{7}77 \\ 12{,}950 \\ \hline 13{,}727 \end{array}$$

⑥ $$\begin{array}{r} {\scriptstyle 2\,4} \\ {\scriptstyle 1\,3} \\ 936 \\ \times\, \boxed{8}5 \\ \hline 4{,}680 \\ 7\boxed{4}{,}880 \\ \hline 79{,}560 \end{array}$$

⑦ $$\begin{array}{r} {\scriptstyle 4} \\ {\scriptstyle 1} \\ 4\boxed{0}8 \\ \times\, 62 \\ \hline 816 \\ 24{,}\boxed{4}80 \\ \hline 25{,}296 \end{array}$$

⑧ $$\begin{array}{r} {\scriptstyle 2\,1} \\ {\scriptstyle 6\,3} \\ \boxed{3}74 \\ \times\, 49 \\ \hline 3{,}366 \\ 1\boxed{4}{,}960 \\ \hline 18{,}326 \end{array}$$

Page 22

① $18 \times 4 \approx 20 \times 4 = 80$ ② $213 \times 8 \approx 200 \times 8 = 1600$

③ $31 \times 9 \approx 30 \times 9 = 270$ ④ $687 \times 3 \approx 700 \times 3 = 2100$

⑤ $49 \times 6 \approx 50 \times 6 = 300$ ⑥ $508 \times 7 \approx 500 \times 7 = 3500$

Note: $50 \times 6 = 300$ because one zero joins to the end of 30.

⑦ $67 \times 8 \approx 70 \times 8 = 560$ ⑧ $819 \times 5 \approx 800 \times 5 = 4000$

Note: $800 \times 5 = 4000$ because two zeroes join to the end of 40.

⑨ $22 \times 3 \approx 20 \times 3 = 60$ ⑩ $994 \times 4 \approx 1000 \times 4 = 4000$

Page 23

① $42 \times 31 \approx 40 \times 30 = 1200$ ② $296 \times 41 \approx 300 \times 40 = 12{,}000$

③ $79 \times 63 \approx 80 \times 60 = 4800$ ④ $811 \times 18 \approx 800 \times 20 = 16{,}000$

⑤ $39 \times 52 \approx 40 \times 50 = 2000$ ⑥ $189 \times 49 \approx 200 \times 50 = 10{,}000$

Note: $40 \times 50 = 2000$ because two zeroes join to the end of 20.

Note: $200 \times 50 = 10{,}000$ because three zeroes join to the end of 10.

⑦ $63 \times 87 \approx 60 \times 90 = 5400$ ⑧ $504 \times 78 \approx 500 \times 80 = 40{,}000$

Note: $500 \times 80 = 40{,}000$ because three zeroes join to the end of 40.

⑨ $98 \times 29 \approx 100 \times 30 = 3000$ ⑩ $413 \times 88 \approx 400 \times 90 = 36{,}000$

Note: $100 \times 30 = 3000$ because three zeroes join to the end of 3.

Page 24

① Since 1 dozen = 12, multiply 12 eggs by 6:

$$\begin{array}{r} {\scriptstyle 1} \\ 12 \\ \times\, 6 \\ \hline 72 \end{array}$$

Sam bought a total of 72 eggs.

② Multiply 73 balls by 8:

$$\begin{array}{r} {\scriptstyle 2} \\ 73 \\ \times\, 8 \\ \hline 584 \end{array}$$

The jars contain a total of 584 balls.

Page 25

③ Multiply $39 by 24:

$$\begin{array}{r} {\scriptstyle 1} \\ {\scriptstyle 3} \\ 39 \\ \times\, 24 \\ \hline 156 \\ 780 \\ \hline 936 \end{array}$$

Alex earned $936 all together.

④ Multiply 61 by 5 and multiply 26 by 7. Then add 305 to 182.

$$\begin{array}{r} 61 \\ \times\, 5 \\ \hline 305 \end{array} \qquad \begin{array}{r} {\scriptstyle 4} \\ 26 \\ \times\, 7 \\ \hline 182 \end{array} \qquad \begin{array}{r} 305 \\ +182 \\ \hline 487 \end{array}$$

There are 487 magnets all together.

⑤ Multiply 48 by 14 and multiply 36 by 25. Then add 672 to 900.

$$\begin{array}{r} \scriptstyle 3 \\ 48 \\ \times\ 14 \\ \hline 192 \\ 480 \\ \hline 672 \end{array} \qquad \begin{array}{r} \scriptstyle 1 \\ \scriptstyle 3 \\ 36 \\ \times\ 25 \\ \hline 180 \\ 720 \\ \hline 900 \end{array} \qquad \begin{array}{r} 900 \\ +672 \\ \hline 1572 \end{array}$$

The total number of blocks is 1572.

Page 26

① $8 \times (5 + 4) = 8 \times 9 = 72$

② $(12 - 8) \times 6 = 4 \times 6 = 24$

③ $3 \times (15 - 8) = 3 \times 7 = 21$

④ $(2 + 3 + 4) \times 6 = 9 \times 6 = 54$

⑤ $(5 + 2) \times (6 + 4) = 7 \times 10 = 70$

⑥ $(3 + 3) \times (9 - 5) = 6 \times 4 = 24$

⑦ $(14 - 6) \times (6 - 2) = 8 \times 4 = 32$

⑧ $(13 - 8) \times (7 + 1) = 5 \times 8 = 40$

Page 27

① $4 \times [2 + (12 \div 3)] = 4 \times [2 + 4] = 4 \times 6 = 24$

② $[(5 \times 3) - 8] \times 7 = [15 - 8] \times 7 = 7 \times 7 = 49$

③ $(14 - 9) \times [(32 \div 8) + 4] = 5 \times [4 + 4] = 5 \times 8 = 40$

④ $8 \times [(7 \times 3) - (6 \times 2)] = 8 \times [21 - 12] = 8 \times 9 = 72$

⑤ $[(9 \times 3) - 20] \times [30 - (4 \times 6)] = [27 - 20] \times [30 - 24] = 7 \times 6 = 42$

Page 28

① $8 + 7 \times 6 = 8 + 42 = 50$ (multiply first)

② $5 \times 4 - 3 = 20 - 3 = 17$ (multiply first)

③ $15 - 6 \times 2 = 15 - 12 = 3$ (multiply first)

④ $4 \times 7 - 2 \times 8 = 28 - 16 = 12$ (multiply first)

⑤ $5 + 4 \times 2 + 7 = 5 + 8 + 7 = 20$ (multiply first)

⑥ $180 \div 3 - 3 \times 4 \times 5 = 60 - 12 \times 5 = 60 - 60 = 0$ (multiply/divide first)

Page 30

① $7 \times 63 = 7 \times (60 + 3) = 7 \times 60 + 7 \times 3 = 420 + 21 = 441$

② $4 \times 71 = 4 \times (70 + 1) = 4 \times 70 + 4 \times 1 = 280 + 4 = 284$

③ $9 \times 82 = 9 \times (80 + 2) = 9 \times 80 + 9 \times 2 = 720 + 18 = 738$

④ $2 \times 39 = 2 \times (40 - 1) = 2 \times 40 - 2 \times 1 = 80 - 2 = 78$

Alternate: $2 \times 39 = 2 \times (30 + 9) = 2 \times 30 + 2 \times 9 = 60 + 18 = 78$

⑤ $6 \times 88 = 6 \times (90 - 2) = 6 \times 90 - 6 \times 2 = 540 - 12 = 528$

Alternate: $6 \times 88 = 6 \times (80 + 8) = 6 \times 80 + 6 \times 8 = 480 + 48 = 528$

⑥ $46 \times 8 = (40 + 6) \times 8 = 40 \times 8 + 6 \times 8 = 320 + 48 = 368$

Alternate: $46 \times 8 = (50 - 4) \times 8 = 50 \times 8 - 4 \times 8 = 400 - 32 = 368$

⑦ $57 \times 5 = (50 + 7) \times 5 = 50 \times 5 + 7 \times 5 = 250 + 35 = 285$

Alternate: $57 \times 5 = (60 - 3) \times 5 = 60 \times 5 - 3 \times 5 = 300 - 15 = 285$

⑧ $99 \times 9 = (100 - 1) \times 9 = 100 \times 9 - 1 \times 9 = 900 - 9 = 891$

Alternate: $99 \times 9 = (90 + 9) \times 9 = 90 \times 9 + 9 \times 9 = 810 + 81 = 891$

Page 31

① (A) $7 \times 6 = 42$

② (C) $9 \times 60 = 540$

③ (C) $40 \times 700 = 28{,}000$ (three zeroes following 28)

④ (E) $87 \times 6 = 522$

⑤ (D) $493 \times 7 = 3451$

⑥ (C) $74 \times 48 = 3552$

⑦ (E) $736 \times 25 = 18{,}400$

⑧ (A) $9\boxed{4}2 \times 8 = 7536$

⑨ (C) $2\boxed{8}5 \times 37 = 10{,}\boxed{5}45$

Page 32

⑩ (B) $71 \times 4 \approx 70 \times 4 = 280$

⑪ (E) $61 \times 39 \approx 60 \times 40 = 2400$ (two zeroes following 24)

⑫ (B) $6 \times 73 = 438$

⑬ (E) $5 \times 32 + 4 \times 24 = 160 + 96 = 256$

⑭ (A) $20 - 6 \times 3 = 20 - 18 = 2$ (multiply first)

⑮ (C) $8 \times [(9 \times 6) - (7 \times 7)] = 8 \times [54 - 49] = 8 \times 5 = 40$

⑯ (B) $7 + 3 \times 2 = 7 + 6 = 13$ (multiply first)

⑰ (B) $9 \times 5 = 5 \times 9$ illustrates the commutative property of multiplication

⑱ (A) $2 \times (3 \times 4) = (2 \times 3) \times 4$ illustrates the associative property of multiplication

⑲ (B) $67 \times 4 = (60 + 7) \times 4 = (60 \times 4) + (7 \times 4)$ according to the distributive property (it may help to review Sec. 1.13)

Note: $67 \times 4 = 268$ and $(60 \times 4) + (7 \times 4) = 240 + 28 = 268$

Chapter 2: Division

Page 34

① 4	② 6	③ 8
④ 9	⑤ 7	⑥ 5
⑦ 4	⑧ 5	⑨ 4
⑩ 9	⑪ 5	⑫ 8
⑬ 6	⑭ 6	⑮ 8
⑯ 9	⑰ 4	⑱ 6
⑲ 3	⑳ 6	㉑ 4
㉒ 8	㉓ 3	㉔ 7
㉕ 7	㉖ 5	㉗ 9
㉘ 9	㉙ 7	㉚ 8

Page 36

①
$$\begin{array}{r} 14 \\ 6\overline{)84} \\ \underline{6} \\ 24 \end{array}$$

②
$$\begin{array}{r} 72 \\ 8\overline{)576} \\ \underline{56} \\ 16 \end{array}$$

Page 37

③
$$\begin{array}{r} 461 \\ 2\overline{)922} \\ \underline{8} \\ 12 \\ \underline{12} \\ 02 \end{array}$$

④
$$\begin{array}{r} 63 \\ 5\overline{)315} \\ \underline{30} \\ 15 \end{array}$$

⑤
$$\begin{array}{r} 49 \\ 7\overline{)343} \\ \underline{28} \\ 63 \end{array}$$

⑥
$$\begin{array}{r} 1428 \\ 3\overline{)4284} \\ \underline{3} \\ 12 \\ \underline{12} \\ 08 \\ \underline{6} \\ 24 \end{array}$$

⑦
```
    238
 4)952
   8
   15
   12
    32
```

⑧
```
    572
 9)5148
   45
    64
    63
     18
```

Page 41

① Only the answers to parts (C) and (D) are given.

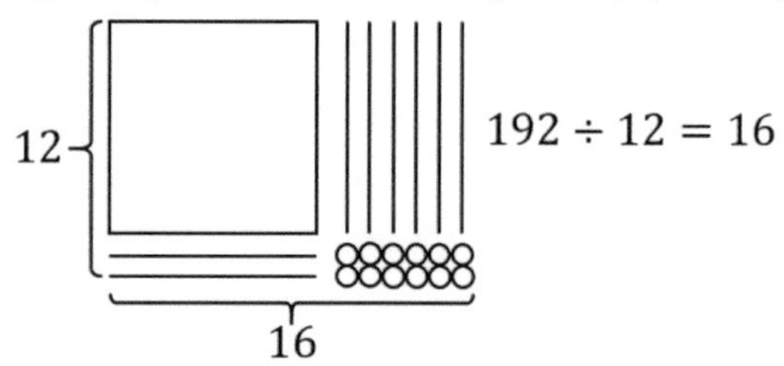

Page 42

② Only the answers to parts (C) and (D) are given.

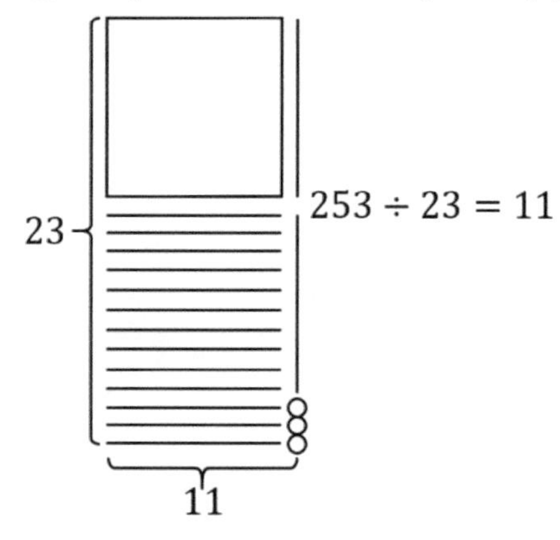

Page 44

①
```
      7
 14)98
    98
     0
```

②
```
      27
 32)864
    64
    224
```

Page 45

③ $$\begin{array}{r} 14 \\ 51\overline{)714} \\ \underline{51} \\ 204 \end{array}$$

④ $$\begin{array}{r} 42 \\ 23\overline{)966} \\ \underline{92} \\ 46 \end{array}$$

⑤ $$\begin{array}{r} 11 \\ 87\overline{)957} \\ \underline{87} \\ 87 \end{array}$$

⑥ $$\begin{array}{r} 17 \\ 43\overline{)731} \\ \underline{43} \\ 301 \end{array}$$

⑦ $$\begin{array}{r} 47 \\ 18\overline{)846} \\ \underline{72} \\ 126 \end{array}$$

⑧ $$\begin{array}{r} 8 \\ 79\overline{)632} \\ \underline{632} \\ 0 \end{array}$$

Page 46

① $$\begin{array}{r} 67 \\ 36\overline{)2412} \\ \underline{216} \\ 252 \end{array}$$

② $$\begin{array}{r} 89 \\ 50\overline{)4450} \\ \underline{400} \\ 450 \end{array}$$

Page 47

③ $$\begin{array}{r} 54 \\ 27\overline{)1458} \\ \underline{135} \\ 108 \end{array}$$

④ $$\begin{array}{r} 62 \\ 96\overline{)5952} \\ \underline{576} \\ 192 \end{array}$$

⑤ $$\begin{array}{r} 92 \\ 18\overline{)1656} \\ \underline{162} \\ 36 \end{array}$$

⑥ $$\begin{array}{r} 34 \\ 65\overline{)2210} \\ \underline{195} \\ 260 \end{array}$$

⑦ $$\begin{array}{r} 77 \\ 41\overline{)3157} \\ \underline{287} \\ 287 \end{array}$$

⑧ $$\begin{array}{r} 48 \\ 83\overline{)3984} \\ \underline{332} \\ 664 \end{array}$$

Page 48

① $$\begin{array}{r} \boxed{7}9 \\ 8\overline{)632} \\ \underline{56} \\ 72 \end{array}$$

② $$\begin{array}{r} 17 \\ 43\overline{)7\boxed{3}1} \\ \underline{43} \\ 301 \end{array}$$

③ $$\begin{array}{r} 573 \\ 6\overline{)3\boxed{4}38} \\ \underline{30} \\ 43 \\ \underline{42} \\ 18 \end{array}$$

Page 50

Partial quotients are on the left. The standard method is on the right.

① $$\begin{array}{rr} 23 & \\ 19\overline{)437} & \\ -\underline{190} & 10 \\ 247 & \\ -\underline{190} & 10 \\ 57 & \\ -\underline{57} & \underline{+3} \\ 0 & 23 \end{array} \qquad \begin{array}{r} 23 \\ 19\overline{)437} \\ \underline{38} \\ 57 \end{array}$$

② $$\begin{array}{rr} 31 & \\ 74\overline{)2294} & \\ -\underline{740} & 10 \\ 1554 & \\ -\underline{740} & 10 \\ 814 & \\ -\underline{740} & 10 \\ 74 & \\ -\underline{74} & \underline{+1} \\ 0 & 31 \end{array} \qquad \begin{array}{r} 31 \\ 74\overline{)2294} \\ \underline{222} \\ 74 \end{array}$$

Page 51

③ $$\begin{array}{rr} 33 & \\ 67\overline{)2211} & \\ -\underline{670} & 10 \\ 1541 & \\ -\underline{670} & 10 \\ 871 & \\ -\underline{670} & 10 \\ 201 & \\ -\underline{201} & \underline{+3} \\ 0 & 33 \end{array} \qquad \begin{array}{r} 33 \\ 67\overline{)2211} \\ \underline{201} \\ 201 \end{array}$$

Partial quotients are on the left. The standard method is on the right.

④

Partial quotients		Standard method
26		26
$43\overline{)1118}$		$43\overline{)1118}$
−430	10	86
688		258
−430	10	
258		
−215	5	
43		
−43	+1	
0	26	

⑤

Partial quotients		Standard method
16		16
$58\overline{)928}$		$58\overline{)928}$
−580	10	58
348		348
−290	5	
58		
−58	+1	
0	16	

⑥

Partial quotients		Standard method
17		17
$81\overline{)1377}$		$81\overline{)1377}$
−810	10	81
567		567
−405	5	
162		
−162	+2	
0	17	

⑦

Partial quotients		Standard method
27		27
$24\overline{)648}$		$24\overline{)648}$
−240	10	48
408		168
−240	10	
168		
−120	5	
48		
−48	+2	
0	27	

⑧

$$\begin{array}{r|r} 41 & \\ 96\overline{)3936} & \\ -\underline{960} & 10 \\ 2976 & \\ -\underline{960} & 10 \\ 2016 & \\ -\underline{960} & 10 \\ 1056 & \\ -\underline{960} & 10 \\ 96 & \\ -\underline{96} & \underline{+1} \\ 0 & 41 \end{array}$$

$$\begin{array}{r} 41 \\ 96\overline{)3936} \\ \underline{384} \\ 96 \end{array}$$

Page 53

① $$\begin{array}{r} 24\ R4 \\ 37\overline{)892} \\ \underline{74} \\ 152 \\ \underline{148} \\ 4 \end{array}$$

② $$\begin{array}{r} 18\ R13 \\ 63\overline{)1147} \\ \underline{63} \\ 517 \\ \underline{504} \\ 13 \end{array}$$

③ $$\begin{array}{r} 67\ R7 \\ 45\overline{)3022} \\ \underline{270} \\ 322 \\ \underline{315} \\ 7 \end{array}$$

④ $$\begin{array}{r} 43\ R1 \\ 92\overline{)3957} \\ \underline{368} \\ 277 \\ \underline{276} \\ 1 \end{array}$$

⑤ $$\begin{array}{r} 35\ R31 \\ 78\overline{)2761} \\ \underline{234} \\ 421 \\ \underline{390} \\ 31 \end{array}$$

⑥ $$\begin{array}{r} 86\ R13 \\ 54\overline{)4657} \\ \underline{432} \\ 337 \\ \underline{324} \\ 13 \end{array}$$

Page 55

① $418 \div 7 \approx 420 \div 7 = 60$ Check: $60 \times 7 = 420$

② $573 \div 8 \approx 560 \div 8 = 70$ Check: $70 \times 8 = 560$

③ $8373 \div 21 \approx 8000 \div 20 = 400$ Check: $400 \times 20 = 8000$

④ $2912 \div 42 \approx 2800 \div 40 = 70$ Check: $70 \times 40 = 2800$

⑤ $2345 \div 29 \approx 2400 \div 30 = 80$ Check: $80 \times 30 = 2400$

⑥ $1536 \div 49 \approx 1500 \div 50 = 30$ Check: $30 \times 50 = 1500$

⑦ $499 \div 72 \approx 490 \div 70 = 7$ Check: $7 \times 70 = 490$

⑧ $4428 \div 88 \approx 4500 \div 90 = 50$ Check: $50 \times 90 = 4500$

⑨ $5294 \div 59 \approx 5400 \div 60 = 90$ Check: $90 \times 60 = 5400$

⑩ $3271 \div 81 \approx 3200 \div 80 = 40$ Check: $40 \times 80 = 3200$

Page 56

① Each student will receive 39 pencils:

$$\begin{array}{r} 39 \\ 32\overline{)1248} \\ \underline{96} \\ 288 \end{array}$$

Check: $32 \times 39 = 1248$

Page 57

② Sally read 24 pages and Patty read 288 pages:

$$\begin{array}{r} 24 \\ 13\overline{)312} \\ \underline{26} \\ 52 \end{array} \qquad \begin{array}{r} 24 \\ \times\, 12 \\ \hline 48 \\ 240 \\ \hline 288 \end{array}$$

Note: Together, they read 13 times as many pages as Sally. (Patty read 12 times as many pages as Sally. When you add Patty's pages and Sally's pages together, this total is 13 times as many as Sally read, since $12 + 1 = 13$.)

Check: Patty read 12 times as many pages as Sally because $12 \times 24 = 288$. The total number of pages read is $24 + 288 = 312$.

③ One children's ticket costs \$6 and one adult ticket costs \$12.

$\$18 \div 3 = \$6$ = the cost of one children's ticket

Tip: The total cost would be the same for 3 children's tickets as it would be for one adult ticket and one children's ticket. This is why the answer is $\$18 \div 3$.

Check: An adult ticket costs twice as much since $\$6 \times 2 = \$12$ and the total cost of one children's ticket and one adult ticket is $\$6 + \$12 = \$18$.

④ One jar has 43 balls (the other 11 jars each have 87 balls)

$$\begin{array}{r} 11\ R43 \\ 87\overline{)1000} \\ \underline{87} \\ 130 \\ \underline{87} \\ 43 \end{array}$$

Divide 1000 by 87 to get 11 $R43$. This shows that 11 jars can be full and that there will be 43 balls leftover to put into the 12th jar.

Check: $87 \times 11 + 43 = 957 + 43 = 1000$

Page 58

① $(24 + 12) \div 9 = 36 \div 9 = 4$

② $30 \div (11 - 6) = 30 \div 5 = 6$

③ $(48 - 16) \div 4 = 32 \div 4 = 8$

④ $35 \div (5 + 4 - 2) = 35 \div (9 - 2) = 35 \div 7 = 5$

⑤ $(9 + 8 + 7) \div 8 = 24 \div 8 = 3$

⑥ $(36 + 36) \div (16 - 8) = 72 \div 8 = 9$

⑦ $(30 - 9) \div (10 - 7) = 21 \div 3 = 7$

⑧ $(24 + 16) \div (3 + 2) = 40 \div 5 = 8$

Page 59

① $[9 + (54 \div 6)] \div 6 = [9 + 9] \div 6 = 18 \div 6 = 3$

② $[(7 \times 8) - 8] \div 8 = [56 - 8] \div 8 = 48 \div 8 = 6$

③ $36 \div [(24 \div 4) + 3] = 36 \div [6 + 3] = 36 \div 9 = 4$

④ $15 \div [(56 \div 7) - (25 \div 5)] = 15 \div [8 - 5] = 15 \div 3 = 5$

⑤ $[(7 \times 6) + 3] \div [13 - (16 \div 2)] = [42 + 3] \div [13 - 8] = 45 \div 5 = 9$

Page 60

① $4 + 6 \div 2 = 4 + 3 = 7$ (divide first)

② $24 \div 6 - 2 = 4 - 2 = 2$ (divide first)

③ $12 - 8 \div 4 = 12 - 2 = 10$ (divide first)

④ $3 \times 8 - 4 \div 2 = 24 - 2 = 22$ (multiply/divide first)

⑤ $9 + 6 \div 3 - 2 = 9 + 2 - 2 = 11 - 2 = 9$ (divide first)

⑥ $9 \times 8 \div 6 - 6 \div 2 = 72 \div 6 - 6 \div 2 = 12 - 3 = 9$ (multiply/divide left to right)

Page 63

① 70 is evenly divisible by 2 and 5 (but not 3, 4, 6, or 9)

Notes: $70 = 2 \times 35$ and $70 = 5 \times 14$

② 78 is evenly divisible by 2, 3, and 6 (but not 4, 5, or 9)

Notes: $78 = 2 \times 39$, $78 = 3 \times 26$, and $78 = 6 \times 13$

③ 220 is evenly divisible by 2, 4, and 5 (but not 3, 6, or 9)

Notes: $220 = 2 \times 110$, $220 = 4 \times 55$, and $220 = 5 \times 44$

④ 195 is evenly divisible by 3 and 5 (but not 2, 4, 6, or 9)

Notes: $195 = 3 \times 65$ and $195 = 5 \times 39$

⑤ 66 is evenly divisible by 2, 3, and 6 (but not 4, 5, or 9)

Notes: $66 = 2 \times 33$, $66 = 3 \times 22$, and $66 = 6 \times 11$

⑥ 72 is evenly divisible by 2, 3, 4, 6, and 9 (but not 5)

Notes: $72 = 2 \times 36$, $72 = 3 \times 24$, $72 = 4 \times 18$, $72 = 6 \times 12$, and $72 = 9 \times 8$

⑦ 180 is evenly divisible by 2, 3, 4, 5, 6, and 9

Notes: $180 = 2 \times 90$, $180 = 3 \times 60$, $180 = 4 \times 45$, $180 = 5 \times 36$, $180 = 6 \times 30$, and $180 = 9 \times 20$

⑧ 495 is evenly divisible by 3, 5, and 9 (but not 2, 4, or 6)

Notes: $495 = 3 \times 165$, $495 = 5 \times 99$, and $495 = 9 \times 55$

⑨ 700 is evenly divisible by 2, 4, and 5 (but not 3, 6, or 9)

Notes: $700 = 2 \times 350$, $700 = 4 \times 175$, and $700 = 5 \times 140$

⑩ 218,790 is evenly divisible by 2, 3, 5, 6, and 9 (but not 4)

Notes: $218{,}790 = 2 \times 109{,}395$; $218{,}790 = 3 \times 72{,}930$; $218{,}790 = 5 \times 43{,}758$; $218{,}790 = 6 \times 36{,}465$; and $218{,}790 = 9 \times 24{,}310$

Page 64

① 4 is the GCF of 8 and 12 ($8 = 4 \times 2$ and $12 = 4 \times 3$)

② 3 is the GCF of 9 and 12 ($9 = 3 \times 3$ and $12 = 3 \times 4$)

③ 6 is the GCF of 24 and 54 ($24 = 6 \times 4$ and $54 = 6 \times 9$)

④ 5 is the GCF of 15 and 20 ($15 = 5 \times 3$ and $20 = 5 \times 4$)

⑤ 9 is the GCF of 36 and 63 ($36 = 9 \times 4$ and $63 = 9 \times 7$)

⑥ 7 is the GCF of 35 and 49 ($35 = 7 \times 5$ and $49 = 7 \times 7$)

Page 65

① 12 is the LCM of 3 and 4 ($3 \times 4 = 12$ and $4 \times 3 = 12$)

② 40 is the LCM of 8 and 10 ($8 \times 5 = 40$ and $10 \times 4 = 40$)

③ 18 is the LCM of 6 and 9 ($6 \times 3 = 18$ and $9 \times 2 = 18$)

④ 50 is the LCM of 10 and 25 ($10 \times 5 = 50$ and $25 \times 2 = 50$)

⑤ 35 is the LCM of 5 and 7 ($5 \times 7 = 35$ and $7 \times 5 = 35$)

⑥ 48 is the LCM of 12 and 16 ($12 \times 4 = 48$ and $16 \times 3 = 48$)

Page 66

① (D) $56 \div 8 = 7$ (note that 49 doesn't evenly divide by 9)

② (C) $320 \div 8 = 40$ Check: $8 \times 40 = 320$

③ (B) $4200 \div 60 = 70$ Check: $60 \times 70 = 4200$ (42 followed by two zeroes)

④ (C) $702 \div 9 = 78$ Check: $9 \times 78 = 702$

⑤ (B) $476 \div 14 = 34$ Check: $14 \times 34 = 476$

⑥ (A) $1537 \div 53 = 29$ Check: $53 \times 29 = 1537$

⑦ (D) 7 because $2\boxed{7}73 \div 47 = 59$ Check: $47 \times 59 = 2773$

⑧ (C) is the best answer $2264 \div 38 \approx 2400 \div 40 = 60$ Check: $40 \times 60 = 2400$

⑨ (C) $R = 3$ because $2325 \div 86 = 27\ R3$ Check: $2322 \div 86 = 27$ and $2325 - 2322 = 3$ (alternatively: $86 \times 27 + 3 = 2322 + 3 = 2325$)

Page 67

⑩ (E) $195 \div 13 = 15$ (1 square + 8 lines + 15 circles = $100 + 80 + 15 = 195$, the height is 13, and the width is 15)

⑪ (A) 15 (the width is 15 because $195 \div 13 = 15$)

⑫ (E) $414 \div 18 = 23$

⑬ (D) Joe is 77 and Amy is 7 Check: Joe (77) is 11 times as old as Amy (7) and their ages add up to $77 + 7 = 84$

⑭ (B) $(72 \div 9) \div 4 = 8 \div 4 = 2$

⑮ (E) $48 \div [(4 \times 3) - (18 \div 3)] = 48 \div [12 - 6] = 48 \div 6 = 8$

⑯ (E) $6 + 12 \div 3 = 6 + 4 = 10$ (divide first)

Page 68

⑰ (C) 90 Check: $90 = 2 \times 45$, $90 = 3 \times 30$, and $90 = 5 \times 18$

⑱ (E) 7614 Check: $7614 = 9 \times 846$ (also, $7 + 6 + 1 + 4 = 18$ is a multiple of 9)

⑲ (C) 8 Check: $32 = 8 \times 4$ and $56 = 8 \times 7$

⑳ (C) 45 Check: $9 \times 5 = 45$ and $15 \times 3 = 45$

Chapter 3: Decimal Place Values

Page 71

① the 2 in 3.2 is in the tenths place
② the 4 in 46 is in the tens place
③ the 5 in 0.25 is in the hundredths place
④ the 8 in 813 is in the hundreds place
⑤ the 7 in 0.007 is in the thousandths place
⑥ the 1 in 1.9 is in the units place (or the ones place)
⑦ the 6 in 2.486 is in the thousandths place
⑧ the 3 in 53,178 is in the thousands place
⑨ the 9 in 5.97 is in the tenths place
⑩ the 0 in 1.9603 is in the thousandths place

Page 72

① $5.3 = 5 + \frac{3}{10}$

② $37.52 = 30 + 7 + \frac{5}{10} + \frac{2}{100}$

Page 73

③ $417.8 = 400 + 10 + 7 + \frac{8}{10}$

④ $0.095 = 0 + \frac{0}{10} + \frac{9}{100} + \frac{5}{1000} = \frac{9}{100} + \frac{5}{1000}$

⑤ $2.837 = 2 + \frac{8}{10} + \frac{3}{100} + \frac{7}{1000}$

⑥ $3812.1 = 3000 + 800 + 10 + 2 + \frac{1}{10}$

⑦ $0.57 = \frac{5}{10} + \frac{7}{100}$

⑧ $29.463 = 20 + 9 + \frac{4}{10} + \frac{6}{100} + \frac{3}{1000}$

⑨ $6.25 = 6 + \frac{2}{10} + \frac{5}{100}$

⑩ $2.5179 = 2 + \frac{5}{10} + \frac{1}{100} + \frac{7}{1000} + \frac{9}{10{,}000}$

Page 75

① 7.2 is seven and two tenths

② 8.14 is eight and fourteen hundredths

③ 94.7 is ninety-four and seven tenths

④ 3.758 is three and seven hundred fifty-eight thousandths

Note: don't put another "and" after the hundred because there is only one decimal point.

⑤ 15.23 is fifteen and twenty-three hundredths

⑥ 8.059 is eight and fifty-nine thousandths

⑦ 423.987 is four hundred twenty-three and nine hundred eighty-seven thousandths (you should only have one "and"; it's for the decimal point)

Page 76

① 5.314

tens	units	.	tenths	hundredths	thousandths
	5	.	3	1	4

② 74.62

tens	units	.	tenths	hundredths	thousandths
7	4	.	6	2	

③ 81.359

tens	units	.	tenths	hundredths	thousandths
8	1	.	3	5	9

Page 79

① Shade 4 out of 10 strips gray for 0.4

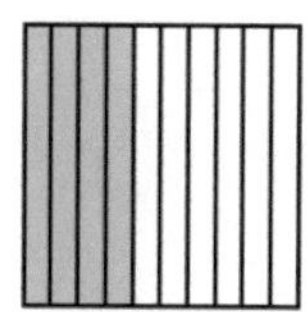

② Shade 6 out of 10 strips gray plus 8 tiny squares for 0.68

③ Shade 3 out of 10 strips gray plus 7 tiny squares plus 4 strips of a magnified tiny square for 0.374

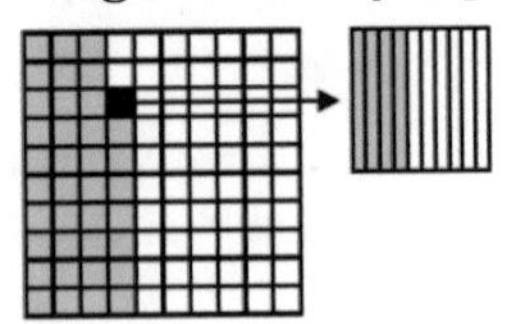

Page 80

① thru ④

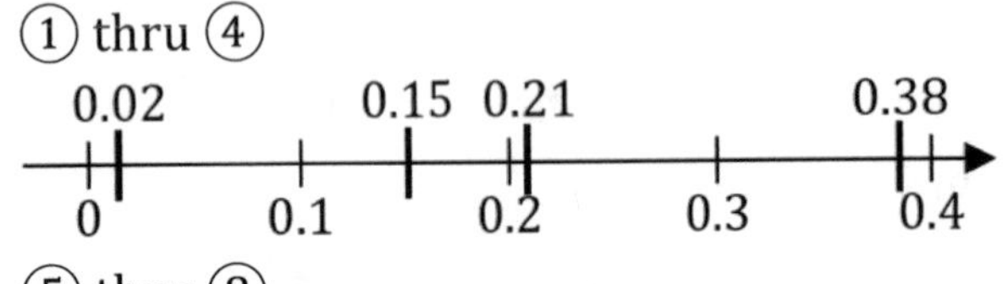

⑤ thru ⑧

0.22 0.47 0.65 0.83

0 0.1 0.2 0.3 0.4 0.5 0.6 0.7 0.8 0.9 1

Page 81

① 0.07 is 10 times larger than 0.007

② 0.04 is 10 times smaller than 0.4

③ 0.9 is 10 times larger than 0.09

④ 0.05 is 10 times smaller than 0.005

⑤ 0.0002 is 10 times smaller than 0.002

⑥ 6 is 10 times larger than 0.6

Page 82

① $0.6 > 0.4$ ② $0.01 < 0.07$ ③ $0.05 = 0.050$

④ $0.3 > 0.03$ ⑤ $0.14 > 0.13$ ⑥ $0.027 < 0.029$

⑦ $0.005 < 0.04$ ⑧ $0.319 < 0.324$ ⑨ $0.634 > 0.633$

Page 83

① 0.18, 0.24, 0.3 Note: 0.3 is equivalent to 0.30

② 0.005, 0.07, 0.3 (thousandths $<$ hundredths $<$ tenths)

③ 0.005, 0.008, 0.009

④ 0.04, 0.07, 0.3 (hundredths $<$ tenths)

⑤ 0.21, 0.216, 0.22 Note: 0.21 and 0.22 are equivalent to 0.210 and 0.220

⑥ 0.69, 0.7, 0.75 Note: 0.7 is equivalent to 0.70

⑦ 0.429, 0.43, 0.435 Note: 0.43 is equivalent to 0.430

Page 85

① 0.3 (the 9 turns the 2 into a 3)
② 0.06 (the 4 leaves the 6 unchanged)
③ 0.7 (the 1 leaves the 7 unchanged)
④ 0.18 (the 8 turns the 7 into an 8)
⑤ 0.1 (the 4 leaves the 1 unchanged)
⑥ 0.35 (the 1 leaves the 5 unchanged)
⑦ 0.592 (the 6 turns the 1 into a 2)
⑧ 0.3 (the 5 turns the 2 into a 3)
⑨ 0.038 (the 4 leaves the 8 unchanged)
⑩ 0.06 (the 5 turns the 5 into a 6)
⑪ 0.1 (the 7 turns the 0 into a 1)
⑫ 0.010 (the 9 increases the 09 to 10)
⑬ 0.01 (the 8 turns the 0 into a 1)
⑭ 0.007 (the 7 turns the 6 into a 7)

Page 86

① (B) In 31.498, the 4 is in the tenths place
② (D) In 0.7583, the 8 is in the thousandths place
③ (C) In 6.4271, the 2 is in the hundredths place
④ (C) $50 + 4 + \frac{9}{100} = 50 + 4 + \frac{0}{10} + \frac{9}{100} = 54.09$
Note: there are 0 tenths (since no tenths were given)
⑤ (D) 0.37 is thirty-seven hundredths (the final digit is in the hundredths place)
⑥ (C) 0.415 is four hundred fifteen thousandths (the final digit is in the thousandths place)
⑦ (B) 0.012 is twelve thousandths (the final digit is in the thousandths place)
⑧ (A) 6.49

Page 87

⑨ (E) 0.8 (8 out of 10 strips)
⑩ (B) 0.46 (4 out of 10 strips plus 6 tiny squares)
⑪ (B) 0.657 (6 out of 10 strips plus 5 tiny squares plus 7 out of 10 strips in the magnified tiny square)
⑫ (C) 0.58 (it is closer to 0.6 than 0.5 since it is greater than 0.55)

Page 88

⑬ (C) $0.8 = 0.08 \times 10$

⑭ (C) $0.003 = 0.03 \div 10$

⑮ (C) 0.4 is equal to 0.40

⑯ (E) 0.003, 0.01, 0.2 (thousandths < hundredths < tenths)

⑰ (B) 0.49, 0.5, 0.51 Note that 0.5 is equivalent to 0.50

⑱ (E) 0.7 (the 5 turns the 6 into a 7)

⑲ (A) 0.10 (the 6 turns the 9 into a 10)

⑳ (D) 0.255 (the 7 turns the 4 into a 5)

Chapter 4: Arithmetic with Decimals

Page 91

①
$$\begin{array}{r} 0.53 \\ +\,0.45 \\ \hline 0.98 \end{array}$$

②
$$\begin{array}{r} {\scriptstyle 1} \\ 2.5 \\ +\,0.7 \\ \hline 3.2 \end{array}$$

③
$$\begin{array}{r} 0.08 \\ +\,0.41 \\ \hline 0.49 \end{array}$$

④
$$\begin{array}{r} {\scriptstyle 1} \\ 0.80 \\ +\,0.41 \\ \hline 1.21 \end{array}$$

⑤
$$\begin{array}{r} {\scriptstyle 1} \\ 1.75 \\ +\,0.50 \\ \hline 2.25 \end{array}$$

⑥
$$\begin{array}{r} {\scriptstyle 1} \\ 0.374 \\ +\,0.562 \\ \hline 0.936 \end{array}$$

⑦
$$\begin{array}{r} {\scriptstyle 1\,1} \\ 24.70 \\ +\,6.93 \\ \hline 31.63 \end{array}$$

⑧
$$\begin{array}{r} {\scriptstyle 1\;\;1} \\ 0.679 \\ +\,0.540 \\ \hline 1.219 \end{array}$$

⑨
$$\begin{array}{r} {\scriptstyle 1} \\ 1.180 \\ +\,0.343 \\ \hline 1.523 \end{array}$$

⑩
$$\begin{array}{r} {\scriptstyle 1} \\ 0.060 \\ +\,0.193 \\ \hline 0.253 \end{array}$$

Page 93

①
$$\begin{array}{r} 3.8 \\ -\,2.3 \\ \hline 1.5 \end{array}$$

②
$$\begin{array}{r} {\scriptstyle 6\,10} \\ 0.\cancel{7}\cancel{0} \\ -\,0.18 \\ \hline 0.52 \end{array}$$

③
$$\begin{array}{r} 0.65 \\ -\,0.03 \\ \hline 0.62 \end{array}$$

④
$$\begin{array}{r} 0.65 \\ -\,0.30 \\ \hline 0.35 \end{array}$$

Note: 0.52 is equivalent to 0.520, which is why the zero isn't shown in the solution to Problem 5 below.

⑤

$$\begin{array}{r} 0.872 \\ -\,0.352 \\ \hline 0.52 \end{array}$$

⑥

$$\begin{array}{r} \scriptstyle 8\ 13\,10 \\ \not{9}.\not{4}\not{0} \\ -\,4.78 \\ \hline 4.62 \end{array}$$

⑦

$$\begin{array}{r} \scriptstyle 6\ \ 10 \\ \not{7}.\not{0} \\ -\,2.6 \\ \hline 4.4 \end{array}$$

⑧

$$\begin{array}{r} \scriptstyle 7\,15 \\ 0.\not{8}\not{5}1 \\ -\,0.360 \\ \hline 0.491 \end{array}$$

⑨

$$\begin{array}{r} \scriptstyle 7\,11\,10 \\ 0.0\not{8}\not{2}\not{0} \\ -\,0.0795 \\ \hline 0.0025 \end{array}$$

⑩

$$\begin{array}{r} \scriptstyle 4\ 13\,10 \\ \not{5}.\not{4}\not{0} \\ -\,1.88 \\ \hline 3.52 \end{array}$$

Page 94

① $7.2 + 5.93 \approx 7 + 6 = 13$ (to the nearest unit)

② $0.894 - 0.31 \approx 0.9 - 0.3 = 0.6$ (to the nearest tenth)

③ $0.078 + 0.0493 \approx 0.08 + 0.05 = 0.13$ (to the nearest hundredth)

④ $24.9 - 6.15 \approx 25 - 6 = 19$ (to the nearest unit)

⑤ $0.913 + 0.0781 \approx 0.91 + 0.08 = 0.99$ (to the nearest hundredth)

Note: If you round to the nearest tenth, you get $0.9 + 0.1 = 1.0$ since 0.0781 rounds to 0.1 if you round to the nearest tenth

⑥ $5.23 - 1.9 \approx 5 - 2 = 3$ (to the nearest unit)

Page 96

① $3.14 \times 10 = 31.4$ (shift 1 place to the right)

② $0.0675 \times 100 = 6.75$ (shift 2 places to the right)

③ $0.25 \times 0.1 = 0.025$ (shift 1 place to the left)

④ $0.048 \times 0.01 = 0.00048$ (shift 2 places to the left)

⑤ $0.57 \times 1000 = 570$ (shift 3 places to the right)

⑥ $7.423 \times 0.1 = 0.7423$ (shift 1 place to the left)

⑦ $2.9 \times 100 = 290$ (shift 2 places to the right)

⑧ $34.2 \times 0.001 = 0.0342$ (shift 3 places to the left)

Page 97

⑨ $1.8795 \times 1000 = 1879.5$ (shift 3 places to the right)

⑩ $0.707 \times 0.01 = 0.00707$ (shift 2 places to the left)

⑪ $0.16 \times 0.001 = 0.00016$ (shift 3 places to the left)

⑫ $0.5 \times 100 = 50$ (shift 2 places to the right)

⑬ $0.0673 \times 10 = 0.673$ (shift 1 place to the right)

⑭ $0.053 \times 0.1 = 0.0053$ (shift 1 place to the left)

⑮ $2.6 \times 0.01 = 0.026$ (shift 2 places to the left)

⑯ $0.79 \times 10 = 7.9$ (shift 1 place to the right)

⑰ $6.42 \times 1000 = 6420$ (shift 3 places to the right)

⑱ $8.1 \times 0.001 = 0.0081$ (shift 3 places to the left)

Page 98

①
$$\begin{array}{r} {\scriptstyle 4} \\ 0.47 \\ \times 6 \\ \hline 2.82 \end{array}$$

②
$$\begin{array}{r} {\scriptstyle 4\;7} \\ 8.59 \\ \times 8 \\ \hline 68.72 \end{array}$$

Page 99

③
$$\begin{array}{r} {\scriptstyle 4} \\ 8.6 \\ \times 7 \\ \hline 60.2 \end{array}$$

④
$$\begin{array}{r} {\scriptstyle 1} \\ 0.57 \\ \times 12 \\ \hline 1.14 \\ 5.70 \\ \hline 6.84 \end{array}$$

⑤
$$\begin{array}{r} {\scriptstyle 1} \\ {\scriptstyle 2} \\ 7.4 \\ \times 36 \\ \hline 44.4 \\ 222.0 \\ \hline 266.4 \end{array}$$

⑥
$$\begin{array}{r} {\scriptstyle 1\,3} \\ 0.639 \\ \times 4 \\ \hline 2.556 \end{array}$$

⑦
$$\begin{array}{r} {\scriptstyle 6\,2} \\ 47.3 \\ \times 9 \\ \hline 425.7 \end{array}$$

⑧
$$\begin{array}{r} {\scriptstyle 5\;4} \\ {\scriptstyle 7\;6} \\ 2.88 \\ \times 68 \\ \hline 23.04 \\ 172.80 \\ \hline 195.84 \end{array}$$

Page 100

① $4 \times 7.3 = 4 \times (7 + 0.3) = 4 \times 7 + 4 \times 0.3 = 28 + 1.2 = \boxed{29.2}$

② $6 \times 0.15 = 6 \times (0.1 + 0.05) = 6 \times 0.1 + 6 \times 0.05 = 0.6 + 0.3 = \boxed{0.9}$

Alternative method: First multiply $6 \times 15 = 6 \times (10 + 5) = 6 \times 10 + 6 \times 5 = 60 + 30 = 90$ and then insert a decimal point to match the two decimal places given in 6×0.15 to get 0.90; then remove the trailing zero to get $\boxed{0.9}$

③ $3 \times 8.7 = 3 \times (8 + 0.7) = 3 \times 8 + 3 \times 0.7 = 24 + 2.1 = \boxed{26.1}$

④ $8 \times 0.049 = 8 \times (0.04 + 0.009) = 8 \times 0.04 + 8 \times 0.009 = 0.32 + 0.072 = \boxed{0.392}$

Alternative method: $8 \times 49 = 8 \times (40 + 9) = 8 \times 40 + 8 \times 9 = 320 + 72 = 392$

Match the three decimal places given to get $\boxed{0.392}$

⑤ $9 \times 0.58 = 9 \times (0.5 + 0.08) = 9 \times 0.5 + 9 \times 0.08 = 4.5 + 0.72 = \boxed{5.22}$

Alternative method: $9 \times 58 = 9 \times (50 + 8) = 9 \times 50 + 9 \times 8 = 450 + 72 = 522$

Match the two decimal places given to get $\boxed{5.22}$

⑥ $2 \times 0.097 = 2 \times (0.09 + 0.007) = 2 \times 0.09 + 2 \times 0.007 = 0.18 + 0.014 = \boxed{0.194}$

⑦ $5 \times 0.888 = 5 \times (0.888) = 5 \times 0.8 + 5 \times 0.08 + 5 \times 0.008 = 4 + 0.4 + 0.04 = \boxed{4.44}$ Alternative method: $5 \times 888 = 5 \times (800 + 80 + 8) = 5 \times 800 + 5 \times 80 + 5 \times 8 = 4000 + 400 + 40 = 4440$ Match the three decimal places given to get 4.440 and remove the trailing zero: $\boxed{4.44}$

⑧ $7 \times 6.94 = 7 \times (6.94) = 7 \times 6 + 7 \times 0.9 + 7 \times 0.04 = 42 + 6.3 + 0.28 = \boxed{48.58}$

Alternative method: $7 \times 694 = 7 \times (600 + 90 + 4) = 7 \times 600 + 7 \times 90 + 7 \times 4 = 4200 + 630 + 28 = 4858$ Match the two decimal places given to get $\boxed{48.58}$

Page 101

① $0.8 \times 0.3 = 0.24$

② $0.7 \times 0.5 = 0.35$

③ $0.4 \times 0.04 = 0.016$

④ $0.09 \times 0.06 = 0.0054$

Page 102

⑤ $0.8 \times 0.8 = 0.64$

⑥ $0.06 \times 0.03 = 0.0018$

⑦ $0.7 \times 0.06 = 0.042$

⑧ $0.5 \times 0.2 = 0.10 = 0.1$

⑨ $0.08 \times 0.04 = 0.0032$

⑩ $0.1 \times 0.01 = 0.001$

⑪ $0.6 \times 0.006 = 0.0036$

⑫ $0.09 \times 0.08 = 0.0072$

⑬ $0.8 \times 0.5 = 0.40 = 0.4$

⑭ $0.053 \times 0.1 = 0.0053$

⑮ $0.09 \times 0.7 = 0.063$ ⑯ $0.009 \times 0.3 = 0.0027$

⑰ $0.002 \times 0.004 = 0.000008$ ⑱ $0.09 \times 0.009 = 0.00081$

Page 104

① $7.2 \times 4.3 = 7 \times 4 + 7 \times 0.3 + 0.2 \times 4 + 0.2 \times 0.3 = 28 + 2.1 + 0.8 + 0.06$ $= \boxed{30.96}$ Note that $0.2 \times 0.3 = 0.06$ has two decimal places because 0.2×0.3 has a combined total of two decimal places.

Page 105

② $9.6 \times 7.8 = 9 \times 7 + 9 \times 0.8 + 0.6 \times 7 + 0.6 \times 0.8 = 63 + 7.2 + 4.2 + 0.48$ $= \boxed{74.88}$ Note that $0.6 \times 0.8 = 0.48$ has two decimal places because 0.6×0.8 has a combined total of two decimal places.

③ $8.4 \times 0.53 = 8 \times 0.5 + 8 \times 0.03 + 0.4 \times 0.5 + 0.4 \times 0.03 = 4.0 + 0.24 +$ $0.20 + 0.012 = \boxed{4.452}$ Note that $0.4 \times 0.03 = 0.012$ has three decimal places because 0.4×0.03 has a combined total of three decimal places.

④ $0.42 \times 0.039 = 0.4 \times 0.03 + 0.4 \times 0.009 + 0.02 \times 0.03 + 0.02 \times 0.009$ $= 0.012 + 0.0036 + 0.0006 + 0.00018 = \boxed{0.01638}$ Note that $0.02 \times 0.009 =$ 0.00018 has five decimal places because 0.02×0.009 has a combined total of five decimal places.

⑤ $0.38 \times 0.27 = 0.3 \times 0.2 + 0.3 \times 0.07 + 0.08 \times 0.2 + 0.08 \times 0.07$ $= 0.06 + 0.021 + 0.016 + 0.0056 = \boxed{0.1026}$ Note that $0.08 \times 0.07 = 0.0056$ has four decimal places because 0.08×0.07 has a combined total of four decimal places.

Page 107

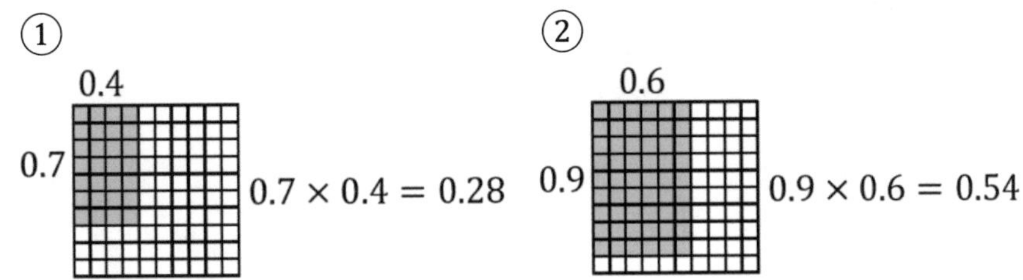

Page 109

①

$$\begin{array}{r} 0.84 \\ \times\ 0.46 \\ \hline \end{array} \rightarrow \begin{array}{r} {\scriptstyle 1} \\ {\scriptstyle 2} \\ 84 \\ \times\ 46 \\ \hline 504 \\ 3360 \\ \hline 3864 \end{array}$$

2 + 2 = 4 decimal places 3864 → $\boxed{0.3864}$

②

$$\begin{array}{r} 5.7 \\ \times\ 0.34 \\ \hline \end{array} \rightarrow \begin{array}{r} {\scriptstyle 2} \\ {\scriptstyle 2} \\ 57 \\ \times\ 34 \\ \hline 228 \\ 1710 \\ \hline 1938 \end{array}$$

1 + 2 = 3 decimal places 1938 → $\boxed{1.938}$

③

$$\begin{array}{r} 6.3 \\ \times\ 2.6 \\ \hline \end{array} \rightarrow \begin{array}{r} {\scriptstyle 1} \\ 63 \\ \times\ 26 \\ \hline 378 \\ 1260 \\ \hline 1638 \end{array}$$

1 + 1 = 2 decimal places 1638 → $\boxed{16.38}$

④

$$\begin{array}{r} 0.77 \\ \times\ 9.2 \\ \hline \end{array} \rightarrow \begin{array}{r} {\scriptstyle 6} \\ {\scriptstyle 1} \\ 77 \\ \times\ 92 \\ \hline 154 \\ 6930 \\ \hline 7084 \end{array}$$

2 + 1 = 3 decimal places 7084 → $\boxed{7.084}$

⑤

$$\begin{array}{r} 0.94 \\ \times\ 0.78 \\ \hline \end{array} \rightarrow \begin{array}{r} {\scriptstyle 2} \\ {\scriptstyle 3} \\ 94 \\ \times\ 78 \\ \hline 752 \\ 6580 \\ \hline 7332 \end{array}$$

2 + 2 = 4 decimal places 7332 → $\boxed{0.7332}$

⑥

$$\begin{array}{r} 0.059 \\ \times\ 0.48 \\ \hline \end{array} \rightarrow \begin{array}{r} {\scriptstyle 3} \\ {\scriptstyle 7} \\ 59 \\ \times\ 48 \\ \hline 472 \\ 2360 \\ \hline 2832 \end{array}$$

3 + 2 = 5 decimal places 2832 → $\boxed{0.02832}$

Page 110

⑦

$$\begin{array}{r} 2.47 \\ \times\ 0.38 \\ \hline \end{array} \rightarrow \begin{array}{r} {\scriptstyle 1\ 2} \\ {\scriptstyle 3\ 5} \\ 247 \\ \times\ 38 \\ \hline 1976 \\ 7410 \\ \hline 9386 \end{array}$$

2 + 2 = 4 decimal places 9386 → $\boxed{0.9386}$

⑧

$$\begin{array}{r} 0.479 \\ \times\ 0.62 \\ \hline \end{array} \rightarrow \begin{array}{r} {\scriptstyle 4\ 5} \\ {\scriptstyle 1\ 1} \\ 479 \\ \times\ 62 \\ \hline 958 \\ 28{,}740 \\ \hline 29{,}698 \end{array}$$

3 + 2 = 5 decimal places 29,698 → $\boxed{0.29698}$

⑨

$$\begin{array}{r} 0.788 \\ \times\ 0.95 \\ \hline \end{array} \rightarrow \begin{array}{r} {\scriptstyle 7\ 7} \\ {\scriptstyle 4\ 4} \\ 788 \\ \times\ 95 \\ \hline 3{,}940 \\ 70{,}920 \\ \hline 74{,}860 \end{array}$$

3 + 2 = 5 decimal places 74,860 → 0.74860 → $\boxed{0.7486}$

⑩

$$\begin{array}{r} 8.36 \\ \times\ 6.7 \\ \hline \end{array} \rightarrow \begin{array}{r} {\scriptstyle 2\ 3} \\ {\scriptstyle 2\ 4} \\ 836 \\ \times\ 67 \\ \hline 5{,}852 \\ 50{,}160 \\ \hline 56{,}012 \end{array}$$

2 + 1 = 3 decimal places 56,012 → $\boxed{56.012}$

⑪

$$\begin{array}{r} 0.693 \\ \times\, 0.84 \\ \hline \end{array} \rightarrow \begin{array}{r} {\scriptstyle 7\,2} \\ {\scriptstyle 3\,1} \\ 693 \\ \times\, 84 \\ \hline 2{,}772 \\ 55{,}440 \\ \hline 58{,}212 \end{array}$$

$3 + 2 = 5$ decimal places $\rightarrow$ $\boxed{0.58212}$

⑫

$$\begin{array}{r} 0.0924 \\ \times\, 0.73 \\ \hline \end{array} \rightarrow \begin{array}{r} {\scriptstyle 1\,2} \\ {\scriptstyle 1} \\ 924 \\ \times\, 73 \\ \hline 2{,}772 \\ 64{,}680 \\ \hline 67{,}452 \end{array}$$

$4 + 2 = 6$ decimal places $\rightarrow$ $\boxed{0.067452}$

Page 111

① $0.713 \times 0.687 \approx 0.7 \times 0.7 = 0.49$

② $1.89 \times 0.72 \approx 2 \times 0.7 = 1.4$

③ $0.794 \times 0.031 \approx 0.8 \times 0.03 = 0.024$

④ $0.406 \times 0.49 \approx 0.4 \times 0.5 = 0.20 = 0.2$

⑤ $7.88 \times 5.96 \approx 8 \times 6 = 48$

⑥ $0.061 \times 0.0807 \approx 0.06 \times 0.08 = 0.0048$

⑦ $19.8 \times 0.42 \approx 20 \times 0.4 = 8$

⑧ $0.098 \times 0.513 \approx 0.10 \times 0.5 = 0.1 \times 0.5 = 0.05$

Page 113

① $0.328 \div 100 = 0.00328$ (shift 2 places to the left)

② $9.2 \div 10 = 0.92$ (shift 1 place to the left)

③ $0.78 \div 0.1 = 7.8$ (shift 1 place to the right)

④ $0.63 \div 0.01 = 63$ (shift 2 places to the right)

⑤ $4.5 \div 1000 = 0.0045$ (shift 3 places to the left)

⑥ $0.8 \div 0.001 = 800$ (shift 3 places to the right)

⑦ $0.397 \div 10 = 0.0397$ (shift 1 place to the left)

⑧ $17 \div 0.1 = 170$ (shift 1 place to the right)

Page 114

⑨ $6.249 \div 100 = 0.06249$ (shift 2 places to the left)

⑩ $0.0381 \div 0.001 = 38.1$ (shift 3 places to the right)

⑪ $0.8324 \div 0.01 = 83.24$ (shift 2 places to the right)

⑫ $0.736 \div 1000 = 0.000736$ (shift 3 places to the left)

⑬ $0.0244 \div 10 = 0.00244$ (shift 1 place to the left)

⑭ $1.09 \div 0.001 = 1090$ (shift 3 places to the right)

⑮ $9.0503 \div 0.01 = 905.03$ (shift 2 places to the right)

⑯ $0.505 \div 100 = 0.00505$ (shift 2 places to the left)

⑰ $0.01 \div 1000 = 0.00001$ (shift 3 places to the left)

⑱ $0.00081 \div 0.1 = 0.0081$ (shift 1 place to the right)

Page 116

①
0.13
$6\overline{)0.78}$
$\underline{0.6}$
0.18

②
3.84
$9\overline{)34.56}$
$\underline{27}$
7.5
$\underline{7.2}$
0.36

③
0.444
$5\overline{)2.22}$
$\underline{2.0}$
0.2
$\underline{0.2}$
0.02

④
0.0625
$8\overline{)0.500}$
$\underline{0.48}$
0.02
$\underline{0.016}$
0.004

⑤
0.0076
$3\overline{)0.0228}$
$\underline{0.021}$
0.0018

⑥
0.0864
$7\overline{)0.6048}$
$\underline{0.56}$
0.044
$\underline{0.042}$
0.0028

Page 117

⑦
0.47
$16\overline{)7.52}$
$\underline{6.4}$
1.12

⑧
0.012
$52\overline{)0.624}$
$\underline{0.52}$
0.104

⑨
0.065
$83\overline{)5.395}$
$\underline{4.98}$
0.415

⑩
0.00075
$44\overline{)0.0330}$
$\underline{0.0308}$
0.0022

⑪
```
     0.004
 75)0.300
    0.300
        0
```

⑫
```
      0.375
 94)35.25
    28.2
     7.05
     6.58
     0.47
```

Page 121

① Only the answers to parts (C) and (D) are given.

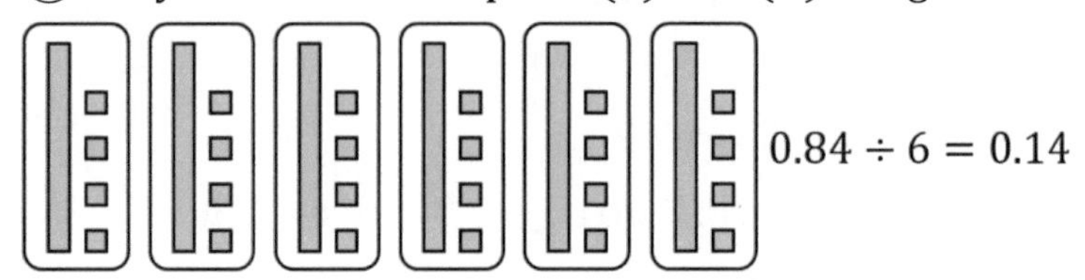

$0.84 \div 6 = 0.14$

Page 122

② Only the answers to parts (C) and (D) are given.

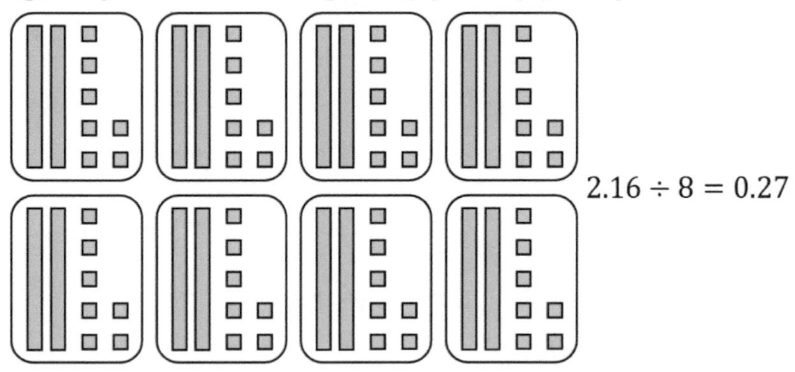

$2.16 \div 8 = 0.27$

Page 123

① $4.092 \div 6 \approx 4.2 \div 6 = 0.7$ Check: $6 \times 0.7 = 4.2$

② $0.394 \div 8 \approx 0.4 \div 8 = 0.05$ Check: $8 \times 0.05 = 0.4$

③ $12.21 \div 5 \approx 12 \div 5 = 2.4$ Check: $5 \times 2.4 = 12$

④ $0.0193 \div 9 \approx 0.018 \div 9 = 0.002$ Check: $9 \times 0.002 = 0.018$

⑤ $3.498 \div 4 \approx 3.6 \div 4 = 0.9$ Check: $4 \times 0.9 = 3.6$

⑥ $5.555 \div 3 \approx 5.4 \div 3 = 1.8$ Check: $3 \times 1.8 = 5.4$

⑦ $0.5 \div 7 \approx 0.49 \div 7 = 0.07$ Check: $7 \times 0.07 = 0.49$

Page 125

① $0.8 + 0.6 = 1.4$ Check: $1.4 - 0.6 = 0.8$

② $0.16 - 0.07 = 0.09$ Check: $0.09 + 0.07 = 0.16$

③ $4 \times 0.9 = 3.6$ Check: $3.6 \div 4 = 0.9$

Note: You may need to show more work (like we did in Sec.'s 4.5, 4.9, and 4.12 for many of these solutions) than we have shown here.

④ $0.42 \div 7 = 0.06$ Check: $7 \times 0.06 = 0.42$

⑤ $1.2 - 0.47 = 0.73$ Check: $0.73 + 0.47 = 1.2$

⑥ $0.62 \times 5 = 3.1$ Check: $3.1 \div 5 = 0.62$

⑦ $0.056 + 0.0056 = 0.0616$ Check: $0.0616 - 0.0056 = 0.056$

⑧ $3.2 \div 8 = 0.4$ Check: $8 \times 0.4 = 3.2$

⑨ $3 - 0.84 = 2.16$ Check: $2.16 + 0.84 = 3$

⑩ $0.054 \div 9 = 0.006$ Check: $9 \times 0.006 = 0.054$

Page 127

① The total weight is 6.35 pounds:

$$\begin{array}{r} {}^{1} \\ 3.72 \\ +\,2.63 \\ \hline 6.35 \end{array}$$

② $3.42 \div 10 = 0.342$ ounces (shift 1 place to the left)

③ Sydney finishes the race in 6.926 seconds:

$$\begin{array}{r} {}^{6\ 12\ 8\ 14} \\ \not{7}.\not{2}\not{9}4 \\ -\,0.368 \\ \hline 6.926 \end{array}$$

Page 128

④ The area of the rectangle is 40.42 square inches:

$$\begin{array}{rcr} & & {}^{2} \\ & & {}^{4} \\ 8.6 & & 86 \\ \times\,4.7 & \rightarrow & \times\,47 \\ \hline & & 602 \\ 1+1=2 & & 3440 \\ \text{decimal places} & & 4042 \\ & \rightarrow & \boxed{40.42} \end{array}$$

⑤ The apples cost \$3.92, the oranges cost \$2.34, and the total cost is \$6.26:

$$\begin{array}{r} \overset{7}{0.49} \\ \times 8 \\ \hline 3.92 \end{array} \qquad \begin{array}{r} \overset{5}{0.39} \\ \times 6 \\ \hline 2.34 \end{array} \qquad \begin{array}{r} \overset{1}{3.92} \\ +2.34 \\ \hline 6.26 \end{array}$$

⑥ Divide 0.78 by 3 to find that there are 0.26 liters of ice. Divide 0.78 by 6 to find that there are 0.13 liters of lemon juice. Add 0.78, 0.26, and 0.13 together to find that the combined volume is 1.17 liters.

$$\begin{array}{r} 0.26 \\ 3\overline{)0.78} \\ \underline{0.6} \\ 0.18 \end{array} \qquad \begin{array}{r} 0.13 \\ 6\overline{)0.78} \\ \underline{0.6} \\ 0.18 \end{array} \qquad \begin{array}{r} \overset{1\ 1}{0.78} \\ 0.26 \\ +0.13 \\ \hline 1.17 \end{array}$$

(Why does the problem mention "by volume" twice in parentheses? Because there are other ways to measure amounts. If it said "by weight," we would need more information to solve the problem because the fraction of the volume and the fraction of the weight may be different, since different substances have different densities.)

Page 129

① $0.2 + 0.8 \times 3 = 0.2 + 2.4 = 2.6$ (multiply first)

② $1.8 \div 3 + 0.4 = 0.6 + 0.4 = 1.0 = 1$ (divide first)

③ $0.3 \times 0.2 - 0.041 = 0.06 - 0.041 = 0.019$ (multiply first)

Note: You may need to show more work (like we did in Sec.'s 4.1, 4.2, 4.9, and 4.12 for many of these solutions) than we have shown here.

④ $0.9 - 0.6 \div 4 = 0.9 - 0.15 = 0.75$ (divide first)

⑤ $(0.36 + 0.54) \times 0.8 = 0.9 \times 0.8 = 0.72$

⑥ $2.8 \div (0.073 - 0.063) = 2.8 \div 0.01 = 280$ (shift 2 places to the right)

⑦ $(0.87 - 0.77) \times (0.38 + 0.46) = 0.1 \times 0.84 = 0.084$ (shift 1 place to the left)

⑧ $(2.5 + 1.7) \div (1 - 0.4) = 4.2 \div 0.6 = 7$ Check: $7 \times 0.6 = 4.2$

Page 130

① (E) 0.842

$$\begin{array}{r} \scriptsize{1\,1} \\ 0.479 \\ +\,0.363 \\ \hline 0.842 \end{array}$$

② (D) 6.76

$$\begin{array}{r} \scriptsize{6\;\;9\,10} \\ \not{7}.\not{0}\not{0} \\ -\,0.24 \\ \hline 6.76 \end{array}$$

③ (C) $0.09106 + 0.06893 \approx 0.09 + 0.07 = 0.16$

④ (E) $0.2934 - 0.1096 \approx 0.3 - 0.1 = 0.2$

⑤ (D) $0.43 \times 100 = 43$ (shift 2 places to the right)

⑥ (A) $0.0579 \times 0.1 = 0.00579$ (shift 1 place to the left)

⑦ (D) 44.73

$$\begin{array}{r} \scriptsize{2\;6} \\ 6.39 \\ \times\,7 \\ \hline 44.73 \end{array}$$

⑧ (B) $0.06 \times 0.005 = 0.00030 = 0.0003$ because $6 \times 5 = 30$ and 30 with 5 decimal places is 0.00030 (then remove the trailing decimal zero)

⑨ (E) $(7 + 0.4) \times (4 + 0.9) = 7 \times 4 + 7 \times 0.9 + 0.4 \times 4 + 0.4 \times 0.9$ $= 28 + 6.3 + 1.6 + 0.36 = 36.26$ (Alternatively, multiply $7.4 \times 4.9 = 36.26$)

Page 131

⑩ (C) $0.8 \times 0.6 = 0.48$

⑪ (E) 67.94

$$\begin{array}{rcr} & & \scriptsize{7} \\ & & \scriptsize{5} \\ 7.9 & & 79 \\ \times\,8.6 & \rightarrow & \times\,86 \\ \hline & & 474 \\ 1 + 1 = 2 & & 6320 \\ \text{decimal places} & & 6794 \\ & \rightarrow & \boxed{67.94} \end{array}$$

⑫ (B) $0.5893 \times 0.3127 \approx 0.6 \times 0.3 = 0.18$ because $6 \times 3 = 18$ and 18 with 2 decimal places is 0.18

⑬ (B) $79.2 \div 1000 = 0.0792$ (shift 3 places to the left)

⑭ (E) $0.0627 \div 0.01 = 6.27$ (shift 2 places to the right)

⑮ (B) 0.0375

$$\begin{array}{r} 0.0375 \\ 4\overline{)0.1500} \\ \underline{0.12} \\ 0.030 \\ \underline{0.028} \\ 0.002 \end{array}$$

⑯ (C) division and multiplication are inverse operations

⑰ (E) 118.5 oz (multiply each pair and then add)

$$\begin{array}{r} {}^{2} \\ 6.3 \\ \times\ 7 \\ \hline 44.1 \end{array} \qquad \begin{array}{r} {}^{3\,2} \\ 18.6 \\ \times\ 4 \\ \hline 74.4 \end{array} \qquad \begin{array}{r} {}^{1} \\ 44.1 \\ +74.4 \\ \hline 118.5 \end{array}$$

Page 132

⑱ (C) each friend receives \$1.66 and there are two pennies left over

Note that \$5 isn't evenly divisible by 3 because the sum of the digits isn't a multiple of 3 (Sec. 2.14). The largest decimal less than \$5 that is evenly divisible by 3 is \$4.98 (because $4 + 9 + 8 = 21$ is a multiple of 3). If we divide \$4.98 by 3, there will be 2 pennies left over because $\$5 - \$4.98 = 0.02$.

$$\begin{array}{r} 1.66 \\ 3\overline{)4.98} \\ \underline{3} \\ 1.9 \\ \underline{1.8} \\ 0.18 \end{array}$$

(Side notes: If the customer paid with a \$5 bill, they will have to get change for the \$5 bill first. The math would be much simpler if they waited until they washed two more cars before dividing the money.)

⑲ (D) $0.7 + 0.3 \times 6 = 0.7 + 1.8 = 2.5$ (multiply first)

⑳ (B) $(0.94 - 0.08) \times 0.2 = 0.86 \times 0.2 = 0.172$

Chapter 5: Arithmetic with Fractions

Page 134

① $\frac{7}{14} = \frac{7 \div 7}{14 \div 7} = \frac{1}{2}$

② $\frac{12}{8} = \frac{12 \div 4}{8 \div 4} = \frac{3}{2}$

③ $\frac{12}{18} = \frac{12 \div 6}{18 \div 6} = \frac{2}{3}$

Note: If you divide by a number that is less than the GCF, you will need at least one more step to complete the problem. For example, if you divide by 3 instead of 6 in Problem 3, you will need two steps instead of one step:

$\frac{12}{18} = \frac{12 \div 3}{18 \div 3} = \frac{4}{6} = \frac{4 \div 2}{6 \div 2} = \frac{2}{3}$

Page 135

④ $\frac{15}{25} = \frac{15 \div 5}{25 \div 5} = \frac{3}{5}$

⑤ $\frac{27}{36} = \frac{27 \div 9}{36 \div 9} = \frac{3}{4}$

⑥ $\frac{16}{20} = \frac{16 \div 4}{20 \div 4} = \frac{4}{5}$

⑦ $\frac{15}{9} = \frac{15 \div 3}{9 \div 3} = \frac{5}{3}$

⑧ $\frac{24}{30} = \frac{24 \div 6}{30 \div 6} = \frac{4}{5}$

⑨ $\frac{28}{16} = \frac{28 \div 4}{16 \div 4} = \frac{7}{4}$

⑩ $\frac{56}{35} = \frac{56 \div 7}{35 \div 7} = \frac{8}{5}$

⑪ $\frac{18}{54} = \frac{18 \div 18}{54 \div 18} = \frac{1}{3}$

⑫ $\frac{32}{72} = \frac{32 \div 8}{72 \div 8} = \frac{4}{9}$

Page 137

① $3\frac{4}{5} = \frac{3 \times 5 + 4}{5} = \frac{15 + 4}{5} = \frac{19}{5}$

② $6\frac{5}{7} = \frac{6 \times 7 + 5}{7} = \frac{42 + 5}{7} = \frac{47}{7}$

③ $4\frac{1}{3} = \frac{4 \times 3 + 1}{3} = \frac{12 + 1}{3} = \frac{13}{3}$

④ $7\frac{3}{4} = \frac{7 \times 4 + 3}{4} = \frac{28 + 3}{4} = \frac{31}{4}$

⑤ $1\frac{2}{3} = \frac{1\times3+2}{3} = \frac{3+2}{3} = \frac{5}{3}$

⑥ $9\frac{7}{8} = \frac{9\times8+7}{8} = \frac{72+7}{8} = \frac{79}{8}$

⑦ $5\frac{1}{6} = \frac{5\times6+1}{6} = \frac{30+1}{6} = \frac{31}{6}$

⑧ $8\frac{3}{8} = \frac{8\times8+3}{8} = \frac{64+3}{8} = \frac{67}{8}$

Page 138

① $\frac{9}{2} = 9 \div 2 = 4\ R1 = 4\frac{1}{2}$ Alternate: $\frac{9}{2} = \frac{8+1}{2} = \frac{8}{2} + \frac{1}{2} = 4\frac{1}{2}$

② $\frac{13}{5} = 13 \div 5 = 2\ R3 = 2\frac{3}{5}$ Alternate: $\frac{13}{5} = \frac{10+3}{5} = \frac{10}{5} + \frac{3}{5} = 2\frac{3}{5}$

③ $\frac{29}{6} = 29 \div 6 = 4\ R5 = 4\frac{5}{6}$ Alternate: $\frac{29}{6} = \frac{24+5}{6} = \frac{24}{6} + \frac{5}{6} = 4\frac{5}{6}$

Page 139

④ $\frac{20}{3} = 20 \div 3 = 6\ R2 = 6\frac{2}{3}$ Alternate: $\frac{20}{3} = \frac{18+2}{3} = \frac{18}{3} + \frac{2}{3} = 6\frac{2}{3}$

⑤ $\frac{37}{8} = 37 \div 8 = 4\ R5 = 4\frac{5}{8}$ Alternate: $\frac{37}{8} = \frac{32+5}{8} = \frac{32}{8} + \frac{5}{8} = 4\frac{5}{8}$

⑥ $\frac{15}{2} = 15 \div 2 = 7\ R1 = 7\frac{1}{2}$ Alternate: $\frac{15}{2} = \frac{14+1}{2} = \frac{14}{2} + \frac{1}{2} = 7\frac{1}{2}$

⑦ $\frac{18}{7} = 18 \div 7 = 2\ R4 = 2\frac{4}{7}$ Alternate: $\frac{18}{7} = \frac{14+4}{7} = \frac{14}{7} + \frac{4}{7} = 2\frac{4}{7}$

⑧ $\frac{25}{3} = 25 \div 3 = 8\ R1 = 8\frac{1}{3}$ Alternate: $\frac{25}{3} = \frac{24+1}{3} = \frac{24}{3} + \frac{1}{3} = 8\frac{1}{3}$

⑨ $\frac{34}{9} = 34 \div 9 = 3\ R7 = 3\frac{7}{9}$ Alternate: $\frac{34}{9} = \frac{27+7}{9} = \frac{27}{9} + \frac{7}{9} = 3\frac{7}{9}$

⑩ $\frac{39}{4} = 39 \div 4 = 9\ R3 = 9\frac{3}{4}$ Alternate: $\frac{39}{4} = \frac{36+3}{4} = \frac{36}{4} + \frac{3}{4} = 9\frac{3}{4}$

⑪ $\frac{59}{8} = 59 \div 8 = 7\ R3 = 7\frac{3}{8}$ Alternate: $\frac{59}{8} = \frac{56+3}{8} = \frac{56}{8} + \frac{3}{8} = 7\frac{3}{8}$

⑫ $\frac{49}{6} = 49 \div 6 = 8\ R1 = 8\frac{1}{6}$ Alternate: $\frac{49}{6} = \frac{48+1}{6} = \frac{48}{6} + \frac{1}{6} = 8\frac{1}{6}$

Page 142

① $\frac{1}{2} = \frac{1\times3}{2\times3} = \frac{3}{6}$ and $\frac{2}{3} = \frac{2\times2}{3\times2} = \frac{4}{6}$

② $\frac{5}{6} = \frac{5\times3}{6\times3} = \frac{15}{18}$ and $\frac{4}{9} = \frac{4\times2}{9\times2} = \frac{8}{18}$

③ $\frac{5}{8} = \frac{5\times5}{8\times5} = \frac{25}{40}$ and $\frac{7}{10} = \frac{7\times4}{10\times4} = \frac{28}{40}$

④ $\frac{4}{3} = \frac{4\times4}{3\times4} = \frac{16}{12}$ and $\frac{5}{4} = \frac{5\times3}{4\times3} = \frac{15}{12}$

⑤ $\frac{7}{12} = \frac{7\times4}{12\times4} = \frac{28}{48}$ and $\frac{3}{16} = \frac{3\times3}{16\times3} = \frac{9}{48}$

⑥ $\frac{6}{7} = \frac{6\times8}{7\times8} = \frac{48}{56}$ and $\frac{1}{8} = \frac{1\times7}{8\times7} = \frac{7}{56}$

Page 143

⑦ $\frac{3}{14} = \frac{3\times3}{14\times3} = \frac{9}{42}$ and $\frac{2}{21} = \frac{2\times2}{21\times2} = \frac{4}{42}$

⑧ $\frac{7}{12} = \frac{7\times5}{12\times5} = \frac{35}{60}$ and $\frac{8}{15} = \frac{8\times4}{15\times4} = \frac{32}{60}$

⑨ $\frac{5}{7} = \frac{5\times4}{7\times4} = \frac{20}{28}$ and $\frac{9}{4} = \frac{9\times7}{4\times7} = \frac{63}{28}$

⑩ $\frac{9}{4} = \frac{9\times2}{4\times2} = \frac{18}{8}$ and $\frac{7}{8}$

Note: $\frac{7}{8}$ already has the LCD. (You don't need to multiply $\frac{7}{8}$ by anything.)

⑪ $\frac{8}{15} = \frac{8\times5}{15\times5} = \frac{40}{75}$ and $\frac{9}{25} = \frac{9\times3}{25\times3} = \frac{27}{75}$

⑫ $\frac{7}{24} = \frac{7\times3}{24\times3} = \frac{21}{72}$ and $\frac{5}{18} = \frac{5\times4}{18\times4} = \frac{20}{72}$

Page 145

① $\frac{3}{5} > \frac{4}{7}$ because $\frac{21}{35} > \frac{20}{35}$ since $\frac{3}{5} = \frac{3\times7}{5\times7} = \frac{21}{35}$ and $\frac{4}{7} = \frac{4\times5}{7\times5} = \frac{20}{35}$

② $\frac{7}{6} > \frac{9}{8}$ because $\frac{28}{24} > \frac{27}{24}$ since $\frac{7}{6} = \frac{7\times4}{6\times4} = \frac{28}{24}$ and $\frac{9}{8} = \frac{9\times3}{8\times3} = \frac{27}{24}$

③ $\frac{2}{3} < \frac{3}{4}$ because $\frac{8}{12} < \frac{9}{12}$ since $\frac{2}{3} = \frac{2\times4}{3\times4} = \frac{8}{12}$ and $\frac{3}{4} = \frac{3\times3}{4\times3} = \frac{9}{12}$

④ $\frac{3}{10} > \frac{4}{15}$ because $\frac{9}{30} > \frac{8}{30}$ since $\frac{3}{10} = \frac{3\times3}{10\times3} = \frac{9}{30}$ and $\frac{4}{15} = \frac{4\times2}{15\times2} = \frac{8}{30}$

⑤ $\frac{7}{3} > \frac{13}{6}$ because $\frac{14}{6} > \frac{13}{6}$ since $\frac{7}{3} = \frac{7\times2}{3\times2} = \frac{14}{6}$ Note: $\frac{13}{6}$ already has the LCD

⑥ $\frac{4}{6} = \frac{6}{9}$ because $\frac{12}{18} = \frac{12}{18}$ since $\frac{4}{6} = \frac{4\times3}{6\times3} = \frac{12}{18}$ and $\frac{6}{9} = \frac{6\times2}{9\times2} = \frac{12}{18}$

Note: $\frac{4}{6}$ and $\frac{6}{9}$ each reduce to $\frac{2}{3}$

Page 146

⑦ $5 < \frac{21}{4}$ because $\frac{20}{4} < \frac{21}{4}$ since $\frac{5}{1} = \frac{5\times4}{1\times4} = \frac{20}{4}$

⑧ $\frac{19}{6} > 3$ because $\frac{19}{6} > \frac{18}{6}$ since $\frac{3}{1} = \frac{3\times6}{1\times6} = \frac{18}{6}$

⑨ $8 > \frac{31}{4}$ because $\frac{32}{4} > \frac{31}{4}$ since $\frac{8}{1} = \frac{8\times4}{1\times4} = \frac{32}{4}$

Page 147

⑩ $\frac{5}{8} < \frac{7}{9}$ because $5 \times 9 = 45$ is less than $8 \times 7 = 56$

⑪ $\frac{8}{9} < \frac{9}{10}$ because $8 \times 10 = 80$ is less than $9 \times 9 = 81$

⑫ $\frac{7}{6} > \frac{10}{9}$ because $7 \times 9 = 63$ is greater than $10 \times 6 = 60$

⑬ $\frac{5}{8} > \frac{7}{12}$ because $5 \times 12 = 60$ is greater than $8 \times 7 = 56$

Page 148

① $\frac{2}{3}+\frac{1}{4}=\frac{2\times4}{3\times4}+\frac{1\times3}{4\times3}=\frac{8}{12}+\frac{3}{12}=\frac{8+3}{12}=\frac{11}{12}$

② $\frac{5}{6}+\frac{7}{9}=\frac{5\times3}{6\times3}+\frac{7\times2}{9\times2}=\frac{15}{18}+\frac{14}{18}=\frac{15+14}{18}=\frac{29}{18}$

③ $\frac{4}{7}+\frac{2}{5}=\frac{4\times5}{7\times5}+\frac{2\times7}{5\times7}=\frac{20}{35}+\frac{14}{35}=\frac{20+14}{35}=\frac{34}{35}$

Page 149

④ $\frac{5}{12}+\frac{7}{20}=\frac{5\times5}{12\times5}+\frac{7\times3}{20\times3}=\frac{25}{60}+\frac{21}{60}=\frac{25+21}{60}=\frac{46}{60}=\frac{46\div2}{60\div2}=\frac{23}{30}$

Note: We reduced $\frac{46}{60}$ to $\frac{23}{30}$ using the method from Sec. 5.1.

⑤ $\frac{2}{3}+\frac{1}{9}=\frac{2\times3}{3\times3}+\frac{1}{9}=\frac{6}{9}+\frac{1}{9}=\frac{6+1}{9}=\frac{7}{9}$ Note: $\frac{1}{9}$ already has the LCD.

⑥ $\frac{7}{6}+\frac{9}{8}=\frac{7\times4}{6\times4}+\frac{9\times3}{8\times3}=\frac{28}{24}+\frac{27}{24}=\frac{28+27}{24}=\frac{55}{24}$

⑦ $\frac{3}{4}+\frac{1}{12}=\frac{3\times3}{4\times3}+\frac{1}{12}=\frac{9}{12}+\frac{1}{12}=\frac{9+1}{12}=\frac{10}{12}=\frac{10\div2}{12\div2}=\frac{5}{6}$ Note: $\frac{1}{12}$ already has the LCD.

Note: We reduced $\frac{10}{12}$ to $\frac{5}{6}$ using the method from Sec. 5.1.

⑧ $\frac{5}{6}+\frac{1}{10}=\frac{5\times5}{6\times5}+\frac{1\times3}{10\times3}=\frac{25}{30}+\frac{3}{30}=\frac{25+3}{30}=\frac{28}{30}=\frac{28\div2}{30\div2}=\frac{14}{15}$

Note: We reduced $\frac{28}{30}$ to $\frac{14}{15}$ using the method from Sec. 5.1.

⑨ $\frac{9}{16}+\frac{7}{24}=\frac{9\times3}{16\times3}+\frac{7\times2}{24\times2}=\frac{27}{48}+\frac{14}{48}=\frac{27+14}{48}=\frac{41}{48}$

Page 150

⑩ $\frac{3}{4}-\frac{1}{3}=\frac{3\times3}{4\times3}-\frac{1\times4}{3\times4}=\frac{9}{12}-\frac{4}{12}=\frac{9-4}{12}=\frac{5}{12}$

⑪ $\frac{4}{5}-\frac{2}{7}=\frac{4\times7}{5\times7}-\frac{2\times5}{7\times5}=\frac{28}{35}-\frac{10}{35}=\frac{28-10}{35}=\frac{18}{35}$

⑫ $\frac{5}{3}-\frac{6}{5}=\frac{5\times5}{3\times5}-\frac{6\times3}{5\times3}=\frac{25}{15}-\frac{18}{15}=\frac{25-18}{15}=\frac{7}{15}$

⑬ $\frac{7}{10}-\frac{1}{6}=\frac{7\times3}{10\times3}-\frac{1\times5}{6\times5}=\frac{21}{30}-\frac{5}{30}=\frac{21-5}{30}=\frac{16}{30}=\frac{16\div2}{30\div2}=\frac{8}{15}$

Note: We reduced $\frac{16}{30}$ to $\frac{8}{15}$ using the method from Sec. 5.1.

⑭ $\frac{3}{4}-\frac{1}{2}=\frac{3}{4}-\frac{1\times2}{2\times2}=\frac{3}{4}-\frac{2}{4}=\frac{1}{4}$

⑮ $\frac{7}{18}-\frac{5}{24}=\frac{7\times4}{18\times4}-\frac{5\times3}{24\times3}=\frac{28}{72}-\frac{15}{72}=\frac{28-15}{72}=\frac{13}{72}$

Page 151

⑯ $4+\frac{2}{3}=\frac{4}{1}+\frac{2}{3}=\frac{4\times 3}{1\times 3}+\frac{2}{3}=\frac{12}{3}+\frac{2}{3}=\frac{12+2}{3}=\frac{14}{3}$

⑰ $\frac{15}{4}-2=\frac{15}{4}-\frac{2}{1}=\frac{15}{4}-\frac{2\times 4}{1\times 4}=\frac{15}{4}-\frac{8}{4}=\frac{15-8}{4}=\frac{7}{4}$

⑱ $\frac{8}{5}+7=\frac{8}{5}+\frac{7}{1}=\frac{8}{5}+\frac{7\times 5}{1\times 5}=\frac{8}{5}+\frac{35}{5}=\frac{8+35}{5}=\frac{43}{5}$

⑲ $3-\frac{1}{3}=\frac{3}{1}-\frac{1}{3}=\frac{3\times 3}{1\times 3}-\frac{1}{3}=\frac{9}{3}-\frac{1}{3}=\frac{9-1}{3}=\frac{8}{3}$

⑳ $5+\frac{9}{2}=\frac{5}{1}+\frac{9}{2}=\frac{5\times 2}{1\times 2}+\frac{9}{2}=\frac{10}{2}+\frac{9}{2}=\frac{10+9}{2}=\frac{19}{2}$

Page 154

① The LCM of 2 and 5 is 10. The answer is $\frac{1}{2}+\frac{2}{5}=\frac{9}{10}$.

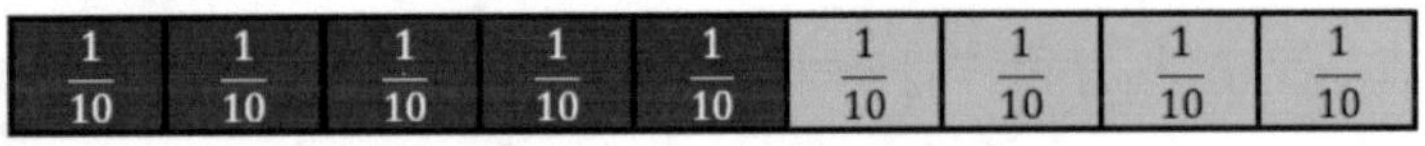

$\frac{1}{10}$	$\frac{1}{10}$	$\frac{1}{10}$	$\frac{1}{10}$	$\frac{1}{10}$	$\frac{1}{10}$	$\frac{1}{10}$	$\frac{1}{10}$	$\frac{1}{10}$

② The LCM of 3 and 4 is 12. The answer is $\frac{2}{3}-\frac{1}{4}=\frac{5}{12}$.

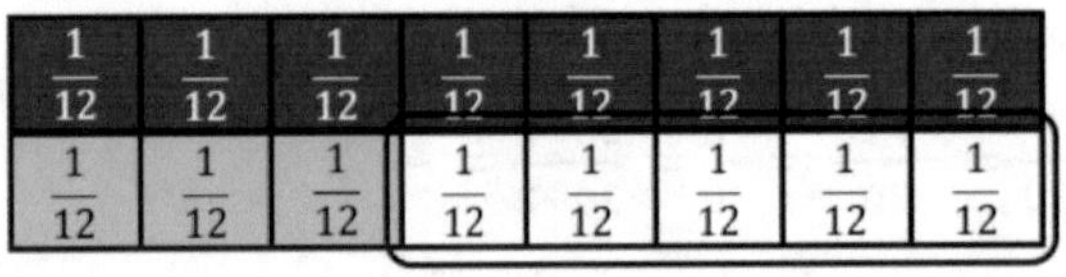

$\frac{1}{12}$	$\frac{1}{12}$	$\frac{1}{12}$	$\frac{1}{12}$	$\frac{1}{12}$	$\frac{1}{12}$	$\frac{1}{12}$	$\frac{1}{12}$
$\frac{1}{12}$	$\frac{1}{12}$	$\frac{1}{12}$	$\frac{1}{12}$	$\frac{1}{12}$	$\frac{1}{12}$	$\frac{1}{12}$	$\frac{1}{12}$

Page 156

① $2\frac{3}{5}+3\frac{1}{4}=2+3+\frac{3\times 4}{5\times 4}+\frac{1\times 5}{4\times 5}=5+\frac{12}{20}+\frac{5}{20}=5+\frac{17}{20}=\boxed{5\frac{17}{20}}$

② $7\frac{1}{2}+7\frac{1}{3}=7+7+\frac{1\times 3}{2\times 3}+\frac{1\times 2}{3\times 2}=14+\frac{3}{6}+\frac{2}{6}=14+\frac{5}{6}=\boxed{14\frac{5}{6}}$

③ $3\frac{5}{6}+2\frac{1}{4}=3+2+\frac{5\times 2}{6\times 2}+\frac{1\times 3}{4\times 3}=5+\frac{10}{12}+\frac{3}{12}=5+\frac{13}{12}=5+1\frac{1}{12}=\boxed{6\frac{1}{12}}$

Note: $\frac{13}{12}=13\div 12=1\,R1=1\frac{1}{12}$

④ $2\frac{3}{4}+1\frac{1}{8}=2+1+\frac{3\times 2}{4\times 2}+\frac{1}{8}=3+\frac{6}{8}+\frac{1}{8}=3+\frac{7}{8}=\boxed{3\frac{7}{8}}$

⑤ $4\frac{2}{3}+7\frac{5}{6}=4+7+\frac{2\times 2}{3\times 2}+\frac{5}{6}=11+\frac{4}{6}+\frac{5}{6}=11+\frac{9}{6}=11+\frac{3}{2}=11+1\frac{1}{2}=\boxed{12\frac{1}{2}}$

Notes: $\frac{9}{6}=\frac{9\div 3}{6\div 3}=\frac{3}{2}$ (we reduced the fraction) and $\frac{3}{2}=3\div 2=1\,R1=1\frac{1}{2}$

⑥ $9\frac{5}{6}+8\frac{7}{9}=9+8+\frac{5\times 3}{6\times 3}+\frac{7\times 2}{9\times 2}=17+\frac{15}{18}+\frac{14}{18}=17+\frac{29}{18}=17+1\frac{11}{18}=\boxed{18\frac{11}{18}}$

Note: $\frac{29}{18}=29\div 18=1\,R11=1\frac{11}{18}$

Page 157

⑦ $4\frac{1}{6}+2\frac{3}{5}=4+2+\frac{1\times5}{6\times5}+\frac{3\times6}{5\times6}=6+\frac{5}{30}+\frac{18}{30}=6+\frac{23}{30}=\boxed{6\frac{23}{30}}$

⑧ $5\frac{2}{3}+6\frac{3}{4}=5+6+\frac{2\times4}{3\times4}+\frac{3\times3}{4\times3}=11+\frac{8}{12}+\frac{9}{12}=11+\frac{17}{12}=11+1\frac{5}{12}=\boxed{12\frac{5}{12}}$

Note: $\frac{17}{12}=17\div12=1\,R5=1\frac{5}{12}$

⑨ $4\frac{1}{4}+7\frac{5}{7}=4+7+\frac{1\times7}{4\times7}+\frac{5\times4}{7\times4}=11+\frac{7}{28}+\frac{20}{28}=11+\frac{27}{28}=\boxed{11\frac{27}{28}}$

⑩ $1\frac{5}{6}+\frac{3}{8}=1+\frac{5\times4}{6\times4}+\frac{3\times3}{8\times3}=1+\frac{20}{24}+\frac{9}{24}=1+\frac{29}{24}=1+1\frac{5}{24}=\boxed{2\frac{5}{24}}$

Note: $\frac{29}{24}=29\div24=1\,R5=1\frac{5}{24}$

⑪ $9\frac{2}{3}+3\frac{7}{9}=9+3+\frac{2\times3}{3\times3}+\frac{7}{9}=12+\frac{6}{9}+\frac{7}{9}=12+\frac{13}{9}=12+1\frac{4}{9}=\boxed{13\frac{4}{9}}$

Note: $\frac{13}{9}=13\div9=1\,R4=1\frac{4}{9}$

⑫ $6\frac{4}{7}+9\frac{3}{8}=6+9+\frac{4\times8}{7\times8}+\frac{3\times7}{8\times7}=15+\frac{32}{56}+\frac{21}{56}=15+\frac{53}{56}=\boxed{15\frac{53}{56}}$

Page 159

① $7\frac{5}{6}-2\frac{3}{4}=(7-2)+\left(\frac{5\times2}{6\times2}-\frac{3\times3}{4\times3}\right)=5+\left(\frac{10}{12}-\frac{9}{12}\right)=5+\frac{1}{12}=\boxed{5\frac{1}{12}}$

② $3\frac{1}{2}-1\frac{2}{3}=2\frac{3}{2}-1\frac{2}{3}=(2-1)+\left(\frac{3\times3}{2\times3}-\frac{2\times2}{3\times2}\right)=1+\left(\frac{9}{6}-\frac{4}{6}\right)=1+\frac{5}{6}=\boxed{1\frac{5}{6}}$

Note: $3\frac{1}{2}=2+1+\frac{1}{2}=2+1\frac{1}{2}=2+\frac{1\times2+1}{2}=2+\frac{3}{2}=2\frac{3}{2}$

Page 160

③ $1\frac{3}{4}-1\frac{3}{8}=(1-1)+\left(\frac{3\times2}{4\times2}-\frac{3}{8}\right)=0+\left(\frac{6}{8}-\frac{3}{8}\right)=0+\frac{3}{8}=\boxed{\frac{3}{8}}$

④ $9\frac{2}{5}-5\frac{4}{7}=8\frac{7}{5}-5\frac{4}{7}=(8-5)+\left(\frac{7\times7}{5\times7}-\frac{4\times5}{7\times5}\right)=3+\left(\frac{49}{35}-\frac{20}{35}\right)=3+\frac{29}{35}=\boxed{3\frac{29}{35}}$

Note: $9\frac{2}{5}=8+1+\frac{2}{5}=8+1\frac{2}{5}=8+\frac{1\times5+2}{5}=8+\frac{7}{5}=8\frac{7}{5}$

⑤ $6\frac{1}{4}-5\frac{1}{3}=5\frac{5}{4}-5\frac{1}{3}=(5-5)+\left(\frac{5\times3}{4\times3}-\frac{1\times4}{3\times4}\right)=0+\left(\frac{15}{12}-\frac{4}{12}\right)=0+\frac{11}{12}=\boxed{\frac{11}{12}}$

Note: $6\frac{1}{4}=5+1+\frac{1}{4}=5+1\frac{1}{4}=5+\frac{1\times4+1}{4}=5+\frac{5}{4}=5\frac{5}{4}$

⑥ $8\frac{4}{9}-4\frac{1}{6}=(8-4)+\left(\frac{4\times2}{9\times2}-\frac{1\times3}{6\times3}\right)=4+\left(\frac{8}{18}-\frac{3}{18}\right)=4+\frac{5}{18}=\boxed{4\frac{5}{18}}$

⑦ $7\frac{2}{7}-\frac{2}{3}=6\frac{9}{7}-\frac{2}{3}=(6-0)+\left(\frac{9\times3}{7\times3}-\frac{2\times7}{3\times7}\right)=6+\left(\frac{27}{21}-\frac{14}{21}\right)=6+\frac{13}{21}=\boxed{6\frac{13}{21}}$

Note: $7\frac{2}{7}=6+1+\frac{2}{7}=6+1\frac{2}{7}=6+\frac{1\times7+2}{7}=6+\frac{9}{7}=6\frac{9}{7}$

⑧ $15\frac{2}{3}-7\frac{5}{6}=14\frac{5}{3}-7\frac{5}{6}=(14-7)+\left(\frac{5\times2}{3\times2}-\frac{5}{6}\right)=7+\left(\frac{10}{6}-\frac{5}{6}\right)=7+\frac{5}{6}=\boxed{7\frac{5}{6}}$

Note: $15\frac{2}{3}=14+1+\frac{2}{3}=14+1\frac{2}{3}=14+\frac{1\times3+2}{3}=14+\frac{5}{3}=14\frac{5}{3}$

Page 161

⑨ $6\frac{2}{5}-1\frac{3}{4}=\frac{6\times5+2}{5}-\frac{1\times4+3}{4}=\frac{32}{5}-\frac{7}{4}=\frac{32\times4}{5\times4}-\frac{7\times5}{4\times5}=\frac{128}{20}-\frac{35}{20}=\frac{93}{20}=\boxed{4\frac{13}{20}}$

⑩ $9\frac{3}{8}-5\frac{7}{12}=\frac{9\times8+3}{8}-\frac{5\times12+7}{12}=\frac{75}{8}-\frac{67}{12}=\frac{75\times3}{8\times3}-\frac{67\times2}{12\times2}=\frac{225}{24}-\frac{134}{24}=\frac{91}{24}=\boxed{3\frac{19}{24}}$

⑪ $12\frac{8}{15}-5\frac{4}{9}=\frac{12\times15+8}{15}-\frac{5\times9+4}{9}=\frac{188}{15}-\frac{49}{9}=\frac{188\times3}{15\times3}-\frac{49\times5}{9\times5}=\frac{564}{45}-\frac{245}{45}=\frac{319}{45}=\boxed{7\frac{4}{45}}$

⑫ $8\frac{1}{6}-2\frac{5}{8}=\frac{8\times6+1}{6}-\frac{2\times8+5}{8}=\frac{49}{6}-\frac{21}{8}=\frac{49\times4}{6\times4}-\frac{21\times3}{8\times3}=\frac{196}{24}-\frac{63}{24}=\frac{133}{24}=\boxed{5\frac{13}{24}}$

Page 163

① $\frac{4}{5}\times2=\frac{4\times2}{5}=\frac{8}{5}$

② $8\times\frac{3}{4}=\frac{8\times3}{4}=\frac{24}{4}=24\div4=6$

③ $9\times\frac{2}{3}=\frac{9\times2}{3}=\frac{18}{3}=18\div3=6$

④ $\frac{7}{6}\times5=\frac{7\times5}{6}=\frac{35}{6}$

⑤ $\frac{4}{9}\times6=\frac{4\times6}{9}=\frac{24}{9}=\frac{24\div3}{9\div3}=\frac{8}{3}$

⑥ $7\times\frac{7}{2}=\frac{7\times7}{2}=\frac{49}{2}$

Page 164

⑦ $8\times\frac{5}{2}=\frac{8\times5}{2}=\frac{40}{2}=40\div2=20$

⑧ $5\times\frac{7}{3}=\frac{5\times7}{3}=\frac{35}{3}$

⑨ $9\times\frac{5}{12}=\frac{9\times5}{12}=\frac{45}{12}=\frac{45\div3}{12\div3}=\frac{15}{4}$

⑩ $6\times\frac{5}{6}=\frac{6\times5}{6}=\frac{30}{6}=30\div6=5$

⑪ $4\times\frac{7}{8}=\frac{4\times7}{8}=\frac{28}{8}=\frac{28\div4}{8\div4}=\frac{7}{2}$

⑫ $7\times\frac{1}{7}=\frac{7\times1}{7}=\frac{7}{7}=7\div7=1$

Page 166

① $\frac{3}{5} \times 20 = 12$

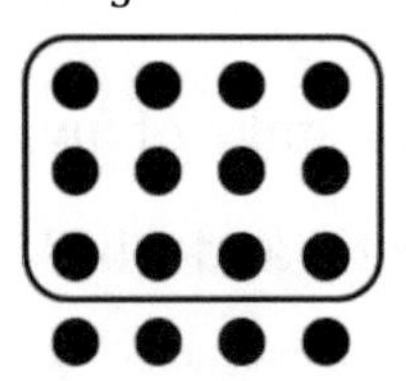

② $18 \times \frac{1}{3} = 6$

③ $\frac{5}{6} \times 30 = 25$

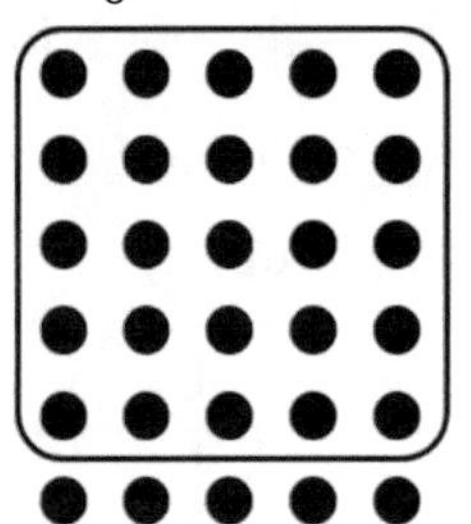

④ $\frac{3}{8} \times 24 = 9$

Page 169

⑤ $\frac{3}{5} \times 2 = \frac{6}{5}$

1					1				
$\frac{1}{5}$	$\frac{1}{5}$	$\frac{1}{5}$	$\frac{1}{5}$	$\frac{1}{5}$	$\frac{1}{5}$	$\frac{1}{5}$	$\frac{1}{5}$	$\frac{1}{5}$	$\frac{1}{5}$

⑥ $4 \times \frac{5}{6} = \frac{10}{3}$ Notes: Divide each strip into 3 pieces, circle 5 out of every 6. There will be 10 circled and each piece is $\frac{1}{3}$, so the answer is $\frac{10}{3}$.

(If you get $\frac{20}{6}$, this reduces to $\frac{10}{3}$. You probably made the mistake of dividing each strip into six pieces instead of three. Since the GCF of 4 and 6 is 2 and since $6 \div 2 = 3$, you should have divided each strip into 3 pieces.)

1			1			1			1		
$\frac{1}{3}$	$\frac{1}{3}$	$\frac{1}{3}$	$\frac{1}{3}$	$\frac{1}{3}$	$\frac{1}{3}$	$\frac{1}{3}$	$\frac{1}{3}$	$\frac{1}{3}$	$\frac{1}{3}$	$\frac{1}{3}$	$\frac{1}{3}$

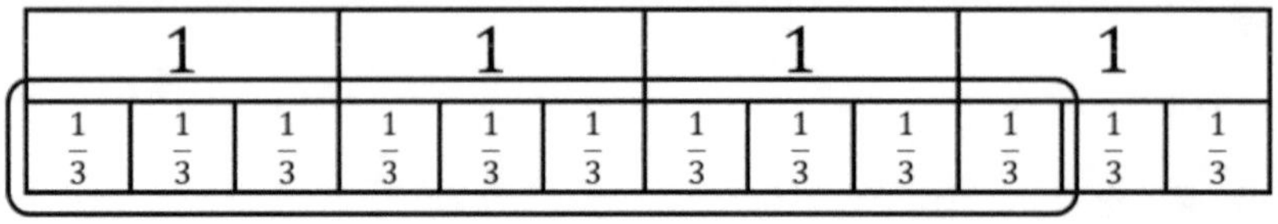

⑦ $\frac{2}{9} \times 6 = \frac{4}{3}$ Notes: Divide each strip into 3 pieces, circle 2 out of every 9. There will be 4 circled and each piece is $\frac{1}{3}$, so the answer is $\frac{4}{3}$.

(If you get $\frac{12}{9}$, this reduces to $\frac{4}{3}$. You probably made the mistake of dividing each strip into nine pieces instead of three. Since the GCF of 6 and 9 is 3 and since $9 \div 3 = 3$, you should have divided each strip into 3 pieces.)

1			1			1			1			1			1		
$\frac{1}{3}$	$\frac{1}{3}$	$\frac{1}{3}$	$\frac{1}{3}$	$\frac{1}{3}$	$\frac{1}{3}$	$\frac{1}{3}$	$\frac{1}{3}$	$\frac{1}{3}$	$\frac{1}{3}$	$\frac{1}{3}$	$\frac{1}{3}$	$\frac{1}{3}$	$\frac{1}{3}$	$\frac{1}{3}$	$\frac{1}{3}$	$\frac{1}{3}$	$\frac{1}{3}$

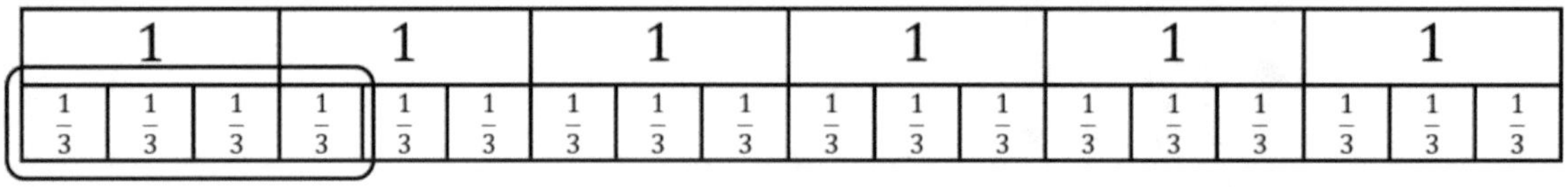

Page 171

① $\frac{3}{4} \div 2 = \frac{3}{4\times2} = \frac{3}{8}$

Tip: Dividing a fraction by a whole number makes a smaller fraction. The fraction is smaller because the numerator is the same while the denominator is larger.

② $5 \div \frac{1}{4} = 5 \times 4 = 20$

Tip: Dividing a whole number by a proper fraction makes a larger whole number.

③ $\frac{1}{4} \div 5 = \frac{1}{4\times5} = \frac{1}{20}$

④ $9 \div \frac{1}{6} = 9 \times 6 = 54$

⑤ $5 \div \frac{1}{2} = 5 \times 2 = 10$

⑥ $\frac{6}{5} \div 9 = \frac{6}{5\times9} = \frac{6}{45} = \frac{6\div3}{45\div3} = \frac{2}{15}$

Note: We reduced $\frac{6}{45}$ to $\frac{2}{15}$ by dividing 6 and 45 each by 3.

Page 172

⑦ $6 \div \frac{1}{8} = 6 \times 8 = 48$

⑧ $\frac{8}{3} \div 4 = \frac{8}{3\times4} = \frac{8}{12} = \frac{8\div4}{12\div4} = \frac{2}{3}$

Note: We reduced $\frac{8}{12}$ to $\frac{2}{3}$ by dividing 8 and 12 each by 4.

⑨ $7 \div \frac{1}{7} = 7 \times 7 = 49$

⑩ $\frac{1}{8} \div 8 = \frac{1}{8\times8} = \frac{1}{64}$

⑪ $\frac{9}{8} \div 6 = \frac{9}{8\times6} = \frac{9}{48} = \frac{9\div3}{48\div3} = \frac{3}{16}$

Note: We reduced $\frac{9}{48}$ to $\frac{3}{16}$ by dividing 9 and 48 each by 3.

⑫ $4 \div \frac{1}{9} = 4 \times 9 = 36$

Page 174

① $2 \div \frac{1}{3} = 6$ When 2 strips are each divided into thirds, there are 6 small pieces.

1			1		
$\frac{1}{3}$	$\frac{1}{3}$	$\frac{1}{3}$	$\frac{1}{3}$	$\frac{1}{3}$	$\frac{1}{3}$

② $\frac{1}{3} \div 2 = \frac{1}{6}$ When one-third is divided by 2, the result is a smaller strip equal to one-sixth.

$\frac{1}{3}$		$\frac{1}{3}$		$\frac{1}{3}$	
$\frac{1}{6}$	$\frac{1}{6}$	$\frac{1}{6}$	$\frac{1}{6}$	$\frac{1}{6}$	$\frac{1}{6}$

Page 176

① $3 \div \frac{1}{4} = 12$ since 3 strips became 12 increments.

Note that $\frac{3}{2} = \frac{3\times2}{2\times2} = \frac{6}{4}$. To find $\frac{3}{2}$ on the number line, go to $\frac{6}{4}$.

② $\frac{1}{4} \div 3 = \frac{1}{12}$ since a single increment had been one-fourth, but is now one-twelfth.

Note that $\frac{1}{2} = \frac{1\times6}{2\times6} = \frac{6}{12}$. To find $\frac{1}{2}$ on the number line, go to $\frac{6}{12}$.

Note that $\frac{5}{6} = \frac{5\times2}{6\times2} = \frac{10}{12}$. To find $\frac{5}{6}$ on the number line, go to $\frac{10}{12}$.

(Since this number line has increments of $\frac{1}{12}$, the idea is to find an equivalent fraction where the denominator equals 12.)

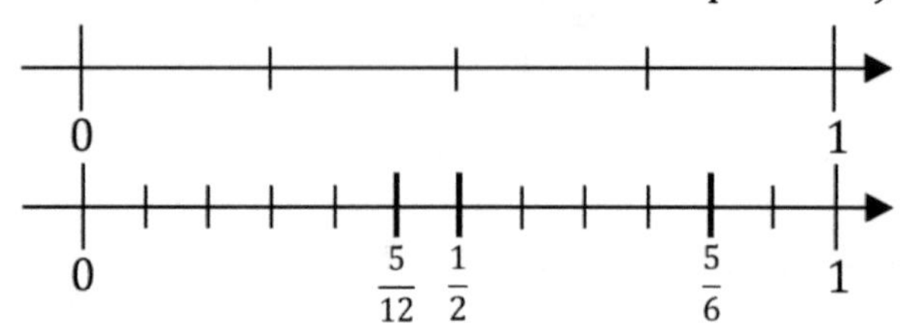

Page 178

① $\frac{6}{11} + \frac{9}{20} \approx \frac{1}{2} + \frac{1}{2} = 1$

② $\frac{10}{9} - \frac{17}{35} \approx 1 - \frac{1}{2} = \frac{1}{2}$

③ $\frac{9}{4} \times \frac{15}{8} \approx 2 \times 2 = 4$

④ $\frac{19}{12} \div \frac{15}{8} \approx \frac{3}{2} \div 2 = \frac{3}{2\times2} = \frac{3}{4}$ (review Sec. 5.11, if needed)

⑤ $\frac{20}{9} - \frac{6}{13} \approx 2 - \frac{1}{2} = \frac{2}{1} - \frac{1}{2} = \frac{2\times2}{1\times2} - \frac{1}{2} = \frac{4}{2} - \frac{1}{2} = \frac{3}{2}$

⑥ $\frac{13}{4} + \frac{9}{5} \approx 3 + 2 = 5$

Page 180

① $\frac{2}{3} - \frac{1}{6} = \frac{2\times2}{3\times2} - \frac{1}{6} = \frac{4}{6} - \frac{1}{6} = \frac{3}{6} = \frac{3\div3}{6\div3} = \boxed{\frac{1}{2}}$ yards

② first step: $\frac{3}{4} + \frac{1}{3} = \frac{3\times3}{4\times3} + \frac{1\times4}{3\times4} = \frac{9}{12} + \frac{4}{12} = \frac{13}{12}$ (**not** the final answer)

second step: $\frac{13}{12} + \frac{5}{8} = \frac{13\times2}{12\times2} + \frac{5\times3}{8\times3} = \frac{26}{24} + \frac{15}{24} = \boxed{\frac{41}{24}}$ cups (equivalent to $1\frac{17}{24}$ cups)

③ $7 \times \frac{3}{4} = \frac{7\times3}{4} = \boxed{\frac{21}{4}}$ pounds (equivalent to $5\frac{1}{4}$ pounds)

Page 181

④ $\frac{3}{4} \div 5 = \frac{3}{4\times5} = \boxed{\frac{3}{20}}$ gallons (using the method from Sec. 5.11)

⑤ Miguel ate $\frac{1}{4} \times 12 = \frac{1\times12}{4} = \frac{12}{4} = 12 \div 4 = \boxed{3}$ slices

Lydia ate $\frac{1}{6} \times 12 = \frac{1\times12}{6} = \frac{12}{6} = 12 \div 6 = \boxed{2}$ slices

Erica ate $12 - 3 - 2 = 12 - 5 = \boxed{7}$ slices

⑥ $2\frac{1}{4} - 1\frac{5}{6} = 1\frac{5}{4} - 1\frac{5}{6} = (1-1) + \left(\frac{5}{4} - \frac{5}{6}\right) = 0 + \left(\frac{5\times3}{4\times3} - \frac{5\times2}{6\times2}\right) = \frac{15}{12} - \frac{10}{12} = \boxed{\frac{5}{12}}$ miles

Note: We used the method from Sec. 5.8.

Alternate solution: $2\frac{1}{4} - 1\frac{5}{6} = \frac{9}{4} - \frac{11}{6} = \frac{9\times3}{4\times3} - \frac{11\times2}{6\times2} = \frac{27}{12} - \frac{22}{12} = \frac{5}{12}$

Page 182

⑦ first step: $24 \times \frac{5}{8} = \frac{24\times5}{8} = \frac{120}{8} = 120 \div 8 = 15$ boys (**not** the final answer)

second step: $24 - 15 = \boxed{9}$ girls

⑧ $5 \div \frac{1}{3} = 5 \times 3 = \boxed{15}$ pieces (using the method from Sec. 5.11)

Alternate solution: There are 3 pieces in each foot: $3 \times 5 = 15$

⑨ first step: $\frac{3}{4} + \frac{1}{3} = \frac{3\times3}{4\times3} + \frac{1\times4}{3\times4} = \frac{9}{12} + \frac{4}{12} = \frac{13}{12}$ (**not** the final answer)

second step: $2\frac{1}{2} - \frac{13}{12} = \frac{5}{2} - \frac{13}{12} = \frac{5\times6}{2\times6} - \frac{13}{12} = \frac{30}{12} - \frac{13}{12} = \frac{17}{12} = \boxed{1\frac{5}{12}}$ hours

Page 183

⑩ $\frac{2}{3} \div 4 = \frac{2}{3\times4} = \frac{2}{12} = \frac{2\div2}{12\div2} = \frac{1}{6}$ feet (using the method from Sec. 5.11)

Note: It is instructive to compare the solutions to Problems 8 and 10. In Problem 8, we counted the number of pieces, whereas in Problem 10 we wish to know how long each piece is.

⑪ first step: $12 \div 4 = 3$ (**not** the final answer) Note: one dozen is equal to 12. It will take 3 times as much flour to make a dozen muffins as it will take to make 4 muffins. Therefore, we need to multiply $\frac{5}{8}$ by 3.

second step $\frac{5}{8} \times 3 = \frac{5\times3}{8} = \boxed{\frac{15}{8}}$ cups (equivalent to $1\frac{7}{8}$ cups)

⑫ first step: $12 \times \frac{2}{3} = \frac{12\times2}{3} = \frac{24}{3} = 24 \div 3 = 8$ pies remain (**not** the final answer)

second step: $12 - 8 = \boxed{4}$ pies were eaten

Since 8 pies were left over, the team must have eaten a total of 4 pies.

Page 184

① (C) $\frac{28\div7}{49\div7}=\frac{4}{7}$

② (E) $4\frac{5}{9}=\frac{4\times9+5}{9}=\frac{36+5}{9}=\frac{41}{9}$

③ (C) $4\frac{5}{6}=\frac{4\times6+5}{6}=\frac{29}{6}$

④ (C) $\frac{1}{24}$ because $24=8\times3$ and $24=12\times2$ (Although 48 and 96 are multiplies of 8 and 12, the lowest multiple of 8 and 12 is 24.)

⑤ (D) $\frac{99}{25}$ because $\frac{11}{3}=\frac{11\times9}{3\times9}=\frac{99}{27}$, not $\frac{99}{25}$

Notes: $\frac{11\times2}{3\times2}=\frac{22}{6}$, $\frac{11\times3}{3\times3}=\frac{33}{9}$, $\frac{11\times7}{3\times7}=\frac{77}{21}$, and $3\frac{2}{3}=\frac{3\times3+2}{3}=\frac{9+2}{3}=\frac{11}{3}$

⑥ (B) $\frac{49}{8}$ is largest. One way to determine this is to express every fraction (except for $\frac{7}{100}$ which is obviously the smallest because it is the only fraction that is smaller than one) with a denominator of 24:

$\frac{49\times3}{8\times3}=\frac{147}{24}$, $\frac{99}{24}$, $4\frac{11}{12}=\frac{4\times12+11}{12}=\frac{59}{12}=\frac{59\times2}{12\times2}=\frac{118}{24}$, and $5\frac{3}{4}=\frac{5\times4+3}{4}=\frac{23}{4}=\frac{23\times6}{4\times6}=\frac{138}{24}$

Since $\frac{147}{24}>\frac{138}{24}>\frac{118}{24}>\frac{99}{24}>\frac{7}{100}$, it follows that $\frac{49}{8}>5\frac{3}{4}>4\frac{11}{12}>\frac{99}{24}>\frac{7}{100}$

⑦ (B) $\frac{9}{4}<2\frac{1}{3}<\frac{17}{6}$ One way to determine this is to express every fraction with a denominator of 12:

$\frac{9}{4}=\frac{9\times3}{4\times3}=\frac{27}{12}$, $2\frac{1}{3}=\frac{2\times3+1}{3}=\frac{6+1}{3}=\frac{7}{3}=\frac{7\times4}{3\times4}=\frac{28}{12}$, and $\frac{17}{6}=\frac{17\times2}{6\times2}=\frac{34}{12}$

Since $\frac{27}{12}<\frac{28}{12}<\frac{34}{12}$, it follows that $\frac{9}{4}<2\frac{1}{3}<\frac{17}{6}$

⑧ (E) $\frac{7}{6}+\frac{2}{9}=\frac{7\times3}{6\times3}+\frac{2\times2}{9\times2}=\frac{21}{18}+\frac{4}{18}=\frac{21+4}{18}=\frac{25}{18}$

Page 185

⑨ (D) $\frac{11}{12}-\frac{1}{6}=\frac{11}{12}-\frac{1\times2}{6\times2}=\frac{11}{12}-\frac{2}{12}=\frac{11-2}{12}=\frac{9}{12}=\frac{9\div3}{12\div3}=\frac{3}{4}$

⑩ (B) $\frac{1}{2}+\frac{1}{2}-\frac{1}{3}=\frac{2}{2}-\frac{1}{3}=1-\frac{1}{3}=\frac{3}{3}-\frac{1}{3}=\frac{3-1}{3}=\frac{2}{3}$

Alternate solution: Visually, the top two strips make 1, so the bottom two strips must also make one. Subtract $\frac{1}{3}$ from 1 to get $\frac{2}{3}$.

⑪ (E) $4\frac{5}{8}+5\frac{3}{4}=4+5+\frac{5}{8}+\frac{3}{4}=9+\frac{5}{8}+\frac{3\times2}{4\times2}=9+\frac{5}{8}+\frac{6}{8}=9+\frac{11}{8}=9+1\frac{3}{8}=10\frac{3}{8}$

⑫ (B) $7\frac{1}{2} - 4\frac{2}{3} = 6\frac{3}{2} - 4\frac{2}{3} = (6-4) + \left(\frac{3}{2} - \frac{2}{3}\right) = 2 + \frac{3\times3}{2\times3} - \frac{2\times2}{3\times2} = 2 + \frac{9}{6} - \frac{4}{6} = 2\frac{5}{6}$

Alternate solution: $7\frac{1}{2} - 4\frac{2}{3} = \frac{15}{2} - \frac{14}{3} = \frac{15\times3}{2\times3} - \frac{14\times2}{3\times2} = \frac{45}{6} - \frac{28}{6} = \frac{17}{6} = 2\frac{5}{6}$

⑬ (D) $24 \times \frac{5}{6} = \frac{24\times5}{6} = \frac{120}{6} = 120 \div 6 = 20$

⑭ (E) $\frac{3}{4} \times 20$

⑮ (E) $12 \div \frac{1}{4} = 12 \times 4 = 48$

Page 186

⑯ (B) $\frac{5}{12} \div 6 = \frac{5}{12\times6} = \frac{5}{72}$

⑰ (C) $\frac{6}{8} = \frac{6\div2}{8\div2} = \frac{3}{4}$

⑱ (C) $\frac{6}{8} = \frac{6\div2}{8\div2} = \frac{3}{4}$ (Coincidentally, Questions 17-18 have the same solution.)

⑲ (B) $\frac{4}{9} \times 36 = \frac{4\times36}{9} = \frac{144}{9} = 16$

⑳ (A) $2\frac{1}{4} - \frac{3}{4} - 1\frac{1}{2} = \frac{9}{4} - \frac{3}{4} - \frac{3}{2} = \frac{6}{4} - \frac{3}{2} = \frac{6}{4} - \frac{3\times2}{2\times2} = \frac{6}{4} - \frac{6}{4} = \frac{0}{4} = 0$

Chapter 6: Data Analysis

Page 189

① Count the frequency for each direction.

Directions that a Car Drove			
N	E	W	S
4	2	5	3

② Count the frequency for each type of ticket.

Number of Tickets Sold		
child	adult	senior
6	2	1

Page 191

① (A) Chihuahua (B) Great Dane (C) $4 + 3 + 5 + 1 + 8 = 21$ dogs (D) 5
(E) Chihuahua

Page 192

② (A) $15 + 60 + 20 = 95$ (B) CD (C) book (D) $20 \times \$12 = \$240$
(E) $15 \times \$5 + 60 \times \$10 + 20 \times \$12 = \$75 + \$600 + \$240 = \$915$

Page 194

① Note: There are other reasonable ways to scale the vertical axis.

Page 195

② Note: There are other reasonable ways to scale the vertical axis.

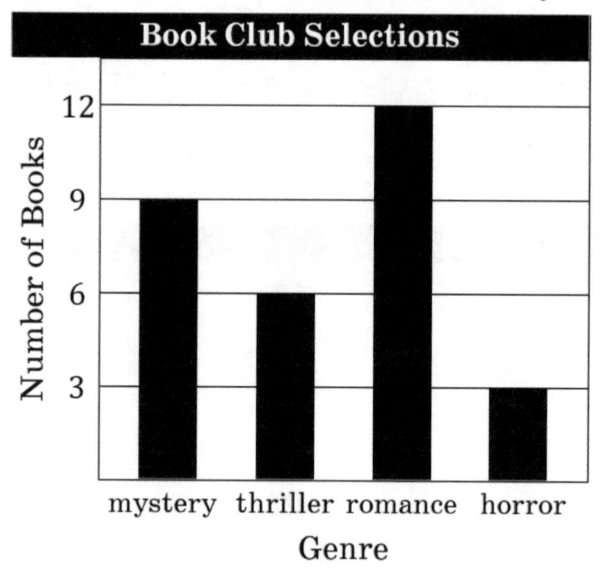

③ Note: There are other reasonable ways to scale the vertical axis.

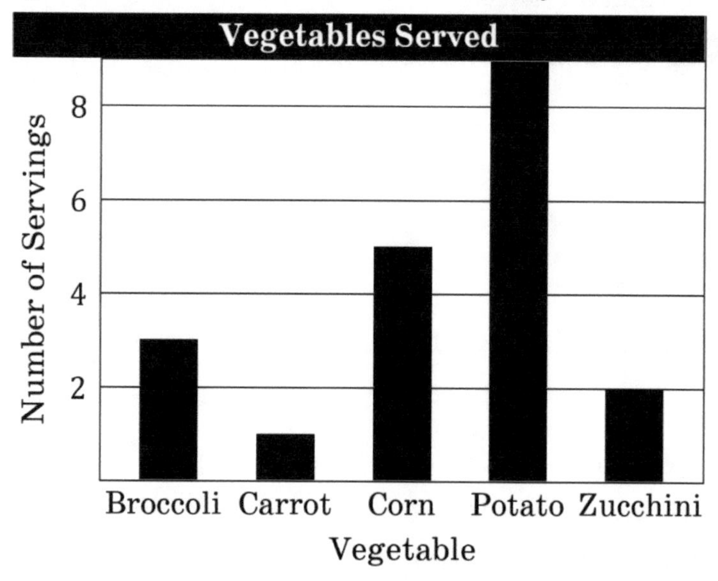

Page 197

① (A) Monday (B) 0.75 (C) $1.75 + 1.5 + 0.75 + 1.25 + 0.25 = 5.5$ hours

(D) $1.5 - 0.75 = 0.75$ hours

Page 198

② (A) $350 \times \$3.25 = \$1137.50$ in the Fall (It may help to review Chapter 4.)

(B) first step: total number of sales = $300 + 150 + 350 + 200 = 1000$ (**not** done yet)

second step: total dollar amount = $1000 \times \$3.25 = \boxed{\$3250}$ (final answer)

(C) first step: difference in number of sales = $300 - 150 = 150$ (**not** done yet)

second step: difference in dollar amount = $150 \times \$3.25 = \boxed{\$487.50}$ (final answer)

Page 200

① There are 21 dots ranging from 1 to 7.

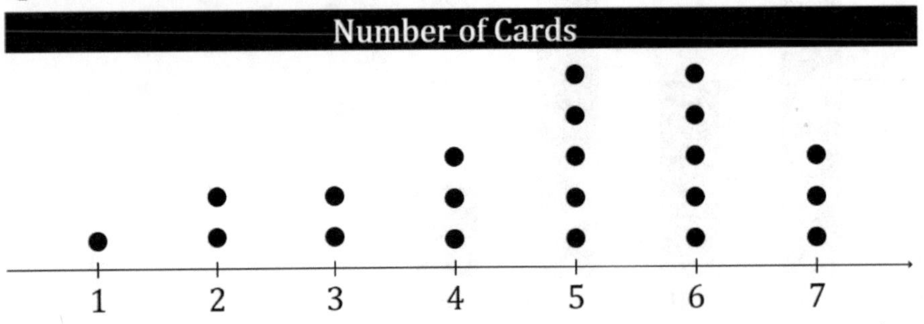

② There are 18 dots ranging from 7 to 9. Be careful with the $7\frac{1}{2}$'s and $8\frac{1}{2}$'s not to count those 7's and 8's twice.

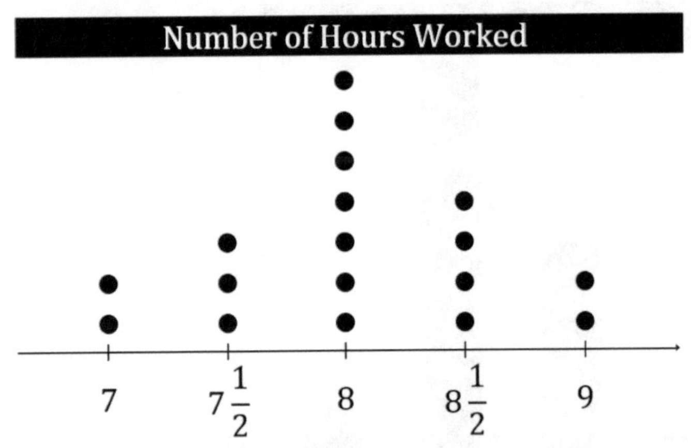

③ There are 13 dots ranging from 11 to 12.5. Be careful with the 11.5's and 12.5's not to count those 11's and 12's twice.

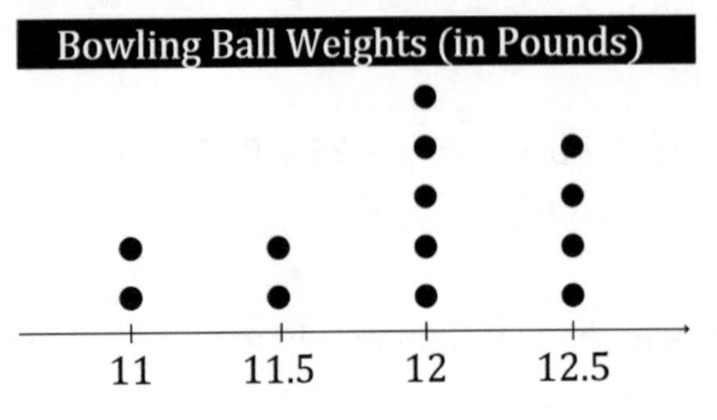

Page 201

① $9 - 3 = 6$

② $24 - 6 = 18$

③ $5 - 5 = 0$

④ $9.3 - 1.7 = 7.6$

⑤ $7\frac{1}{3} - 2\frac{3}{4} = 6\frac{4}{3} - 2\frac{3}{4} = (6-2) + \left(\frac{4}{3} - \frac{3}{4}\right) = 4 + \frac{4\times4}{3\times4} - \frac{3\times3}{4\times3} = 4 + \frac{16}{12} - \frac{9}{12} = 4\frac{7}{12}$

Alternate solution: $7\frac{1}{3} - 2\frac{3}{4} = \frac{22}{3} - \frac{11}{4} = \frac{22\times4}{3\times4} - \frac{11\times3}{4\times3} = \frac{88}{12} - \frac{33}{12} = \frac{55}{12} = 4\frac{7}{12}$

Note: It may help to review Sec. 5.8.

⑥ $0.09 - 0.0375 = 0.0525$ Note: It may help to review Sec. 4.2.

⑦ $\frac{9}{4} - \frac{2}{7} = \frac{9\times7}{4\times7} - \frac{2\times4}{7\times4} = \frac{63}{28} - \frac{8}{28} = \boxed{\frac{55}{28}}$ (equivalent to $1\frac{27}{28}$)

Note that $\frac{9}{4} > \frac{6}{5}$ since $\frac{9}{4} = \frac{9\times5}{4\times5} = \frac{45}{20}$, $\frac{6}{5} = \frac{6\times4}{5\times4} = \frac{24}{20}$, and $\frac{45}{20} > \frac{24}{20}$ (Sec. 5.4)

Note that $\frac{2}{7} < \frac{3}{8}$ since $\frac{2}{7} = \frac{2\times8}{7\times8} = \frac{16}{56}$, $\frac{3}{8} = \frac{3\times7}{8\times7} = \frac{21}{56}$, and $\frac{16}{56} < \frac{21}{56}$

Page 203

① (A) $1\frac{1}{4} - \frac{3}{4} = \frac{5}{4} - \frac{3}{4} = \frac{2}{4} = \frac{1}{2}$ gallon

(B) $3 + 5 + 6 + 5 + 3 = 22$ data points

(C) $3 \times \frac{3}{4} + 5 \times \frac{7}{8} + 6 \times 1 + 5 \times 1\frac{1}{8} + 3 \times 1\frac{1}{4}$ (not finished yet)

Tip: Apply the distributive property (Sec. 1.13) in reverse to regroup:

$3 \times \left(\frac{3}{4} + 1\frac{1}{4}\right) + 5 \times \left(\frac{7}{8} + 1\frac{1}{8}\right) + 6 \times 1 = 3 \times 2 + 5 \times 2 + 6 \times 1 = 6 + 10 + 6 = \boxed{22}$

Note: The answers to (B) and (C) happen to be the same because the average value of the data points happens to be one.

Page 204

② (A) $\$3.25 - \$1.75 = \$1.50$

(B) $7 - 4 = \boxed{3}$ (7 items cost \$2.75 and 4 items cost \$1.75)

(C) $4 + 6 + 6 + 7 + 7 + 6 + 5 = 41$

(D) $4 \times \$1.75 + 6 \times \$2 + 6 \times \$2.25 + 7 \times \$2.5 + 7 \times \$2.75 + 6 \times \$3 + 5 \times \$3.25$
$= \$7 + \$12 + \$13.5 + \$17.5 + \$19.25 + \$18 + \$16.25 = \$103.50$

Page 207

① There are 14 temperatures ranging from the 50's to the 90's.

High Temperatures	
Stems	Leaves
5	4 9
6	0 7 8
7	0 3 4 4
8	0 1 6 8
9	2
5|4 represents 54.	

② There are 14 ages ranging from the 20's to the 60's (but there aren't any ages in the 50's).

Ages at a Book Club	
Stems	Leaves
2	5 7 8
3	0 2 2 2 6 6 8 9
4	1 2
5	
6	5
2|5 represents 25.	

③ There are 11 scores ranging from the 80's to the 110's. Note, for example, that 110 is separated into 11 and 0 (it has 11 tens and 0 ones).

Basketball Scores	
Stems	Leaves
8	9
9	0 2 5 6
10	0 1 3 6
11	0 4
8|9 represents 89.	

Page 209

① (A) The youngest is 7 years old. The oldest is 22 years old. (Maybe the twenty-year old's are adults who work at the camp.)

(B) 12 years is the most common age. (It occurs 6 times.)

(C) 9 people are older than 12. (Their ages range from 13 to 22.)

(D) 31. (There are 12 on top, 15 in the middle, and 4 in the bottom: $12 + 15 + 4 = 31$.)

Page 210

② (A) $75 − $36 = $39 (B) $45 (It occurs 4 times.)

(C) 27 occasions (There are 2 in the top row, 9 in the second row, 8 in the third row, 4 in the fourth row, and 4 in the last row: $2 + 9 + 8 + 4 + 4 = 27$.)

(D) $27 \times \$55 = \$1485 \approx \$1500$

Note: There is more than one way to estimate the answer to Part (D). If your answer is a bit different, it may be okay.

Page 214

① This isn't the only reasonable way to scale the axes. You will probably need to interpolate (read between the lines) to draw some of the points.

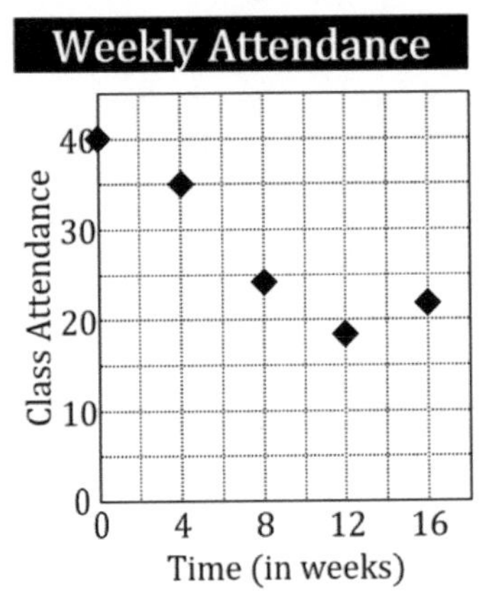

Page 215

② This isn't the only reasonable way to scale the axes. You will probably need to interpolate (read between the lines) to draw some of the points.

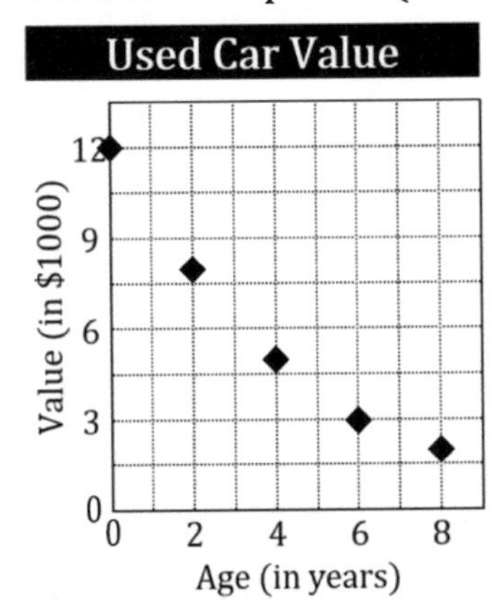

Page 216

③ This isn't the only reasonable way to scale the axes. You will probably need to interpolate (read between the lines) to draw some of the points.

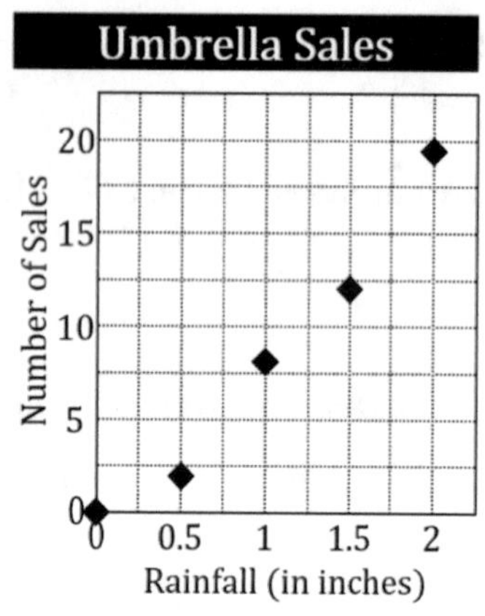

Page 219

① (A) direct proportion (B) $12 (halfway between $8 and $16)
(C) 2.25 hours (note that $20 is halfway between $16 and $24, and that the marker for $18 is halfway between $16 and $20)
(D) $16 (it's between $12 and $18, and closer to $18 since 2 is closer to 2.25 than 2 is to 1.5)

Page 220

② (A) inverse proportion (B) 2 (the third score is exactly 50, which isn't less than 50) (C) 45 (halfway between 40 and 50; note that 50 itself is halfway between 40 and 60) (D) 70 (halfway between 80 and 60)

Page 221

① (D) stem-and-leaf plot (they are all 2 digits in the 60's, 70's, and 80's; also, they aren't repeated enough to suit a dot plot or bar graph well)
② (B) dot plot (there are a handful of numerical values, and most are repeated rather than occurring only once)
③ (A) bar graph (there are categories and frequencies, and the frequencies are rather high for a dot plot)
④ (C) scatter plot (there are two variables in direct proportion)
⑤ (E) time (measured in years); the time is the independent variable (the *height* is *dependent*, not independent, because her height changes as she ages)

Page 222

⑥ (A) January (24)

⑦ (E) $21 + 15 = 36$

⑧ (D) 35 (halfway between 30 and 40)

⑨ (B) $25 - 15 = 10$

Page 223

⑩ (A) $8 - 4 = 4$

⑪ (C) 4 (there are 4 dots above the 6)

⑫ (E) $4 \times 4 + 5 \times 5 + 6 \times 4 + 7 \times 3 + 8 \times 2 = 16 + 25 + 24 + 21 + 16 = 102$

⑬ (B) 24 (occurred three times)

⑭ (C) 3 times (31, 31, and 36)

Page 224

⑮ (D) 0.75 N

⑯ (A) directly proportional

⑰ (C) 0.625 N (greater than 0.5 N and less than 0.75 N; about halfway in between)

⑱ (C) scatter plot

Chapter 7: Number Patterns

Page 227

① $54 + 9 = \boxed{63}, 63 + 9 = \boxed{72}$

② $67 - 8 = \boxed{59}, 27 - 8 = \boxed{19}$

③ $20 + 12 = \boxed{32}, 68 + 12 = \boxed{80}$

④ $159 - 19 = \boxed{140}, 83 - 19 = \boxed{64}$

⑤ $1 + 1 = 2, 2 + 2 = 4, 4 + 3 = \boxed{7}, 7 + 4 = 11, 11 + 5 = 16, 16 + 6 = \boxed{22}$, $22 + 7 = 29$ Note: This pattern adds 1, adds 2, adds 3, adds 4, etc. Each time, the pattern adds one more than last time.

⑥ $8 + 7 = 15, 15 + 9 = 24, 24 + 7 = 31, 31 + 9 = 40, 40 + 7 = \boxed{47}$, $47 + 9 = 56, 56 + 7 = \boxed{63}$ Note: This is an **alternating** pattern. Add 7, then add 9, then add 7, then add 9, then add 7, then add 9, etc.

⑦ $0.25 + 0.125 = \boxed{0.375}, 0.5 + 0.125 = \boxed{0.625}$

⑧ $\frac{7}{16} + \frac{1}{16} = \frac{8}{16} = \boxed{\frac{1}{2}}, \frac{11}{16} + \frac{1}{16} = \frac{12}{16} = \boxed{\frac{3}{4}}$ Note: This pattern is more obvious if you don't reduce the fractions. If you don't reduce the fractions, it looks like this: $\frac{5}{16}, \frac{6}{16}, \frac{7}{16}, \frac{8}{16}, \frac{9}{16}, \frac{10}{16}, \frac{11}{16}, \frac{12}{16}$.

⑨ $4\frac{1}{2} + 1\frac{1}{4} = \boxed{5\frac{3}{4}}, 5\frac{3}{4} + 1\frac{1}{4} = (5 + 1) + \left(\frac{3}{4} + \frac{1}{4}\right) = 6 + 1 = \boxed{7}$

⑩ $5, 4, 5, 5, 5, 6, 5, 7, \boxed{5}, \boxed{8}$ Note: This sequence has two different patterns merged together. One pattern is 5, 5, 5, 5, 5, ... (It adds 0 each time.) The other pattern is 4, 5, 6, 7, 8, ... (It adds 1 each time.) When we merge these two patterns, we get $5, \boxed{4}, 5, \boxed{5}, 5, \boxed{6}, 5, \boxed{7}, 5, \boxed{8}, \ldots$

Page 229

① $54 \times 3 = \boxed{162}, 162 \times 3 = \boxed{486}$

② $4802 \div 7 = \boxed{686}, 98 \div 7 = \boxed{14}$

Tip: If you read this pattern right to left, it's simpler: $2 \times 7 = 14$ and $98 \times 7 = 686$

③ $0.32 \times 4 = \boxed{1.28}, 5.12 \times 4 = \boxed{20.48}$

④ $1000 \div 10 = \boxed{100}, 0.1 \div 10 = \boxed{0.01}$

⑤ $\frac{1}{8} \times 2 = \frac{2}{8} = \boxed{\frac{1}{4}}, 2 \times 2 = \boxed{4}$

⑥ $3 \div 2 = \boxed{\frac{3}{2}}, \frac{3}{2} \div 2 = \boxed{\frac{3}{4}}$

⑦ $4 \times 3 = 12, 12 \times 4 = 48, 48 \times 5 = 240, 240 \times 6 = \boxed{1440}, 1440 \times 7 = \boxed{10{,}080}$

Note: This pattern multiplies by 3, multiplies by 4, multiplies by 5, multiplies by 6, multiplies by 7, etc. Each time, the pattern multiplies by one more than the previous time.

⑧ $\frac{2}{625} \times 5 = \boxed{\frac{2}{125}}, \frac{2}{5} \times 5 = \boxed{2}$ Tip: Since the 2's don't change, just focus on the denominators: 3125, 625, 125, 25, 5, 1. Divide each denominator by 5.

⑨ $\frac{1}{720} \times 6 = \frac{1}{120}, \frac{1}{120} \times 5 = \frac{1}{24}, \frac{1}{24} \times 4 = \boxed{\frac{1}{6}}, \frac{1}{6} \times 3 = \frac{1}{2}, \frac{1}{2} \times 2 = \boxed{1}$

Note: This pattern multiplies by 6, multiplies by 5, multiplies by 4, multiplies by 3, multiplies by 2, etc. Each time, the pattern multiplies by one less than the previous time.

Tip: Just focus on the denominators: 720, 120, 24, 6, 2, 1. It's even simpler if you read them left to right: $1 \times 2 = 2, 2 \times 3 = 6, 6 \times 4 = 24, 24 \times 5 = 120$, $120 \times 6 = 720$ (These numbers have a special name: factorials.)

⑩ The numerators and denominators make two separate patterns:

Divide the numerator by 3 each time: 729, 243, 81, 27, 9, 3.

Divide the denominator by 2 each time: 64, 32, 16, 8, 4, 2.

The missing numbers are $\boxed{\frac{81}{16}}$ and $\boxed{\frac{3}{2}}$.

Page 231

① 47 and 53 (since $45 = 5 \times 9, 49 = 7 \times 7$, and $51 = 3 \times 17$)

② 2 (skip 3), 5 (skip 7), 11 (skip 13), $\boxed{17}$ (skip 19), 23 (skip 29), 31 (skip 37), 41 (skip 43), $\boxed{47}$ Note: This pattern skips every other prime number.

③ 97, 89, 83, $\boxed{79}$, 73, 71, $\boxed{67}$, 61 Note: These prime numbers are in reverse.

Notes: $81 = 9 \times 9, 77 = 7 \times 11, 75 = 5 \times 15, 69 = 3 \times 23, 65 = 5 \times 13$, and $63 = 7 \times 9$

④ $\boxed{35}$ and $\boxed{39}$ Note: These are composite odd numbers. This pattern skips prime numbers (11, 13, 17, 19, 23, 29, 31, and 37 are skipped).

Notes: $35 = 5 \times 7$ and $39 = 3 \times 13$

⑤ $2 - 1 = 1, 3 - 1 = 2, 5 - 1 = 4, 7 - 1 = 6, 11 - 1 = 10, 13 - 1 = 12$, $17 - 1 = \boxed{16}, 19 - 1 = \boxed{18}$ Note: This pattern subtracts 1 from each prime number.

⑥ $2 \times 0.4 = 0.8, 3 \times 0.4 = 1.2, 5 \times 0.4 = 2, 7 \times 0.4 = 2.8, 11 \times 0.4 = \boxed{4.4}$, $13 \times 0.4 = 5.2, 17 \times 0.4 = 6.8, 19 \times 0.4 = \boxed{7.6}$ Note: This pattern multiplies each prime number by 0.4.

⑦ $\frac{2}{3}, \frac{3}{5}, \frac{5}{7}, \frac{7}{11}, \frac{11}{13}, \frac{13}{17}, \boxed{\frac{17}{19}}, \boxed{\frac{19}{23}}$ Notes: The numerators are prime numbers beginning from 2. The denominators are prime numbers beginning from 3.

⑧ $2\frac{3}{4}, 7\frac{11}{12}, 17, \frac{19}{20}, 29\frac{31}{32}, 41\frac{43}{44}, 53\frac{59}{60}, \boxed{67\frac{71}{72}}, \boxed{79\frac{83}{84}}$

Notes: The main pattern is 2, 3 (skip 5), 7, 11 (skip 13), 17, 19 (skip 23), 29, 31 (skip 37), 41, 43 (skip 47), 53, 59 (skip 61), 67, 71 (skip 73), 79, 83. One number is the whole number, the next number is the numerator, a number is skipped, the next number is the whole number, the next number is the numerator, a number is skipped, etc. Each denominator is one more than the corresponding numerator.

Notes: $63 = 7 \times 9, 65 = 5 \times 13, 69 = 3 \times 23, 75 = 5 \times 15, 77 = 7 \times 11$, and $81 = 9 \times 9$

Page 232

① $9 + 12 = \boxed{21}, 54 + 87 = \boxed{141}$

② $0.82 + 1.32 = \boxed{2.14}, 1.32 + 2.14 = \boxed{3.46}$

③ $0.38 + 0.61 = \boxed{0.99}, 0.61 + 0.99 = \boxed{1.6}$

④ $\frac{5}{6} + \frac{4}{3} = \frac{5}{6} + \frac{4 \times 2}{3 \times 2} = \frac{5}{6} + \frac{8}{6} = \boxed{\frac{13}{6}}, \frac{4}{3} + \frac{13}{6} = \frac{4 \times 2}{3 \times 2} + \frac{13}{6} = \frac{8}{6} + \frac{13}{6} = \frac{21}{6} = \boxed{\frac{7}{2}}$

⑤ $5\frac{1}{4} + 8\frac{1}{2} = (5 + 8) + \left(\frac{1}{4} + \frac{1}{2}\right) = \boxed{13\frac{3}{4}}, 8\frac{1}{2} + 13\frac{3}{4} = (8 + 13) + \left(\frac{1}{2} + \frac{3}{4}\right) =$ $21 + 1\frac{1}{4} = \boxed{22\frac{1}{4}}$

Page 234

① $C = A + 9$ (one way to determine this is $16 - 7 = 9$)

A	7	14	21	28	35
C	16	23	30	37	44

② $t = h + 1.645$ (one way to determine this is $2.02 - 0.375 = 1.645$)

Notes: Add 0.375 going across the top row. For example, $0.375 + 0.375 = 0.75$.

h	0.375	0.75	1.125	1.5	1.875
t	2.02	2.395	2.77	3.145	3.52

③ $M = L - \frac{1}{12}$, which is equivalent to $L = M + \frac{1}{12}$ (one way to determine this is $\frac{1}{3} - \frac{1}{4} = \frac{1\times4}{3\times4} - \frac{1\times3}{4\times3} = \frac{4}{12} - \frac{3}{12} = \frac{1}{12}$) Notes: Add $\frac{1}{6}$ going across the top. Note that the top is equivalent to $\frac{2}{6}, \frac{3}{6}, \frac{4}{6}, \frac{5}{6}, \frac{6}{6}$ except that the answers in the table are reduced. The bottom row is equivalent to $\frac{3}{12}, \frac{5}{12}, \frac{7}{12}, \frac{9}{12}, \frac{11}{12}$ except for reducing answers.

L	$\frac{1}{3}$	$\frac{1}{2}$	$\frac{2}{3}$	$\frac{5}{6}$	1
M	$\frac{1}{4}$	$\frac{5}{12}$	$\frac{7}{12}$	$\frac{3}{4}$	$\frac{11}{12}$

Page 235

④ $V = 89 + 12 = \boxed{101}$ when $U = 89$

U	9	17	25	33	41
V	21	29	37	45	53

⑤ $L = 5.11 - 0.27 = \boxed{4.84}$ when $K = 5.11$

K	0.7	1.33	1.96	2.59	3.22
L	0.43	1.06	1.69	2.32	2.95

⑥ $Q = 9\frac{1}{3} + 2\frac{1}{6} = 11\frac{3}{6} = \boxed{11\frac{1}{2}}$ when $P = 9\frac{1}{3}$ Note: $\frac{1}{3} + \frac{1}{6} = \frac{2}{6} + \frac{1}{6} = \frac{3}{6} = \frac{1}{2}$

P	$2\frac{2}{3}$	4	$5\frac{1}{3}$	$6\frac{2}{3}$	8
Q	$4\frac{5}{6}$	$6\frac{1}{6}$	$7\frac{1}{2}$	$8\frac{5}{6}$	$10\frac{1}{6}$

Page 237

① $e = 7b$ For example, $63 = 7 \times 9$ Note: Add 9 going across the top row.

b	9	18	27	36	45
e	63	126	189	252	315

② $d = 5t$ For example, $2.1 = 5 \times 0.42$ Note: Add 0.42 going across the top row.

t	0.42	0.84	1.26	1.68	2.1
d	2.1	4.2	6.3	8.4	10.5

③ $Q = 6C$ For example, $3 = 6 \times \frac{1}{2}$ (since $\frac{6}{2} = 6 \div 2 = 3$)

Note: Add $\frac{1}{4}$ going across the top row. The top row is equivalent to $\frac{2}{4}, \frac{3}{4}, \frac{4}{4}, \frac{5}{4}, \frac{6}{4}$ except that the answers in the table are reduced. The bottom row is equivalent to $\frac{6}{2}, \frac{9}{2}, \frac{12}{2}, \frac{15}{2}, \frac{18}{2}$ except for reducing answers.

C	$\frac{1}{2}$	$\frac{3}{4}$	1	$\frac{5}{4}$	$\frac{3}{2}$
Q	3	$\frac{9}{2}$	6	$\frac{15}{2}$	9

Page 238

④ $v = 12 \times 73 = \boxed{876}$ when $a = 73$

a	8	9	10	11	12
v	96	108	120	132	144

⑤ $P = 0.8 \times 29 = \boxed{23.2}$ when $V = 29$

V	5	7	9	11	13
P	4	5.6	7.2	8.8	10.4

⑥ $y = \frac{2}{3} \times 48 = \frac{96}{3} = 96 \div 3 = \boxed{32}$ when $x = 48$

x	6	8	11	15	20
y	4	$\frac{16}{3}$	$\frac{22}{3}$	10	$\frac{40}{3}$

Page 241

① Plot each (x, y) pair and then draw a line through the points. Each value of y is larger than the corresponding value of x by 3. The slope is 1: The graph below has a rise of 6 and a run of 6. The y-intercept is 3: The line intersects the y-axis at $y = 3$. The equation for this line is $y = x + 3$.

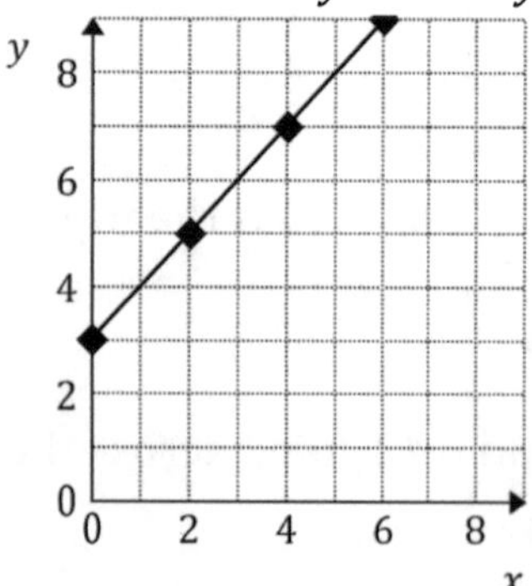

② Plot each (x, y) pair and then draw a line through the points. Each value of y is *smaller* than the corresponding value of x by 2. The slope is 1: The graph below has a rise of 7 and a run of 7. The y-intercept is -2: The line would intersect the y-axis at $y = -2$ (but you can't see this below because we didn't plot any negative values; you can see the x-intercept, but that's different from the y-intercept). The equation for this line is $y = x - 2$.

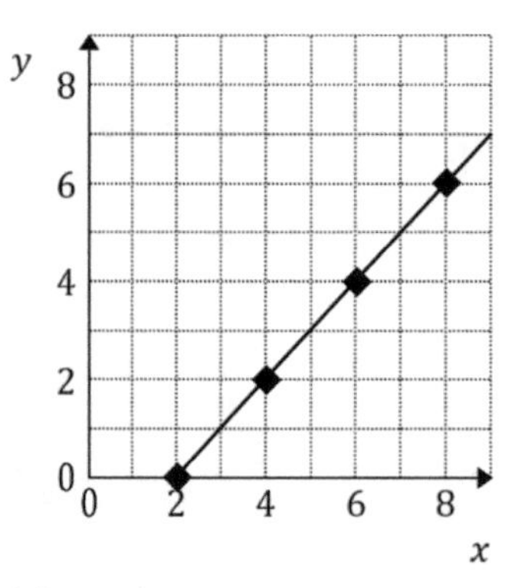

Page 244

① Plot each (x, y) pair and then draw a line through the points. Each value of y is 3 times larger than the corresponding value of x. The graph below has a rise of 9 and a run of 3. The slope is $\frac{9}{3} = 3$. The y-intercept is 0: The line passes through the origin. The equation for this line is $y = 3x$.

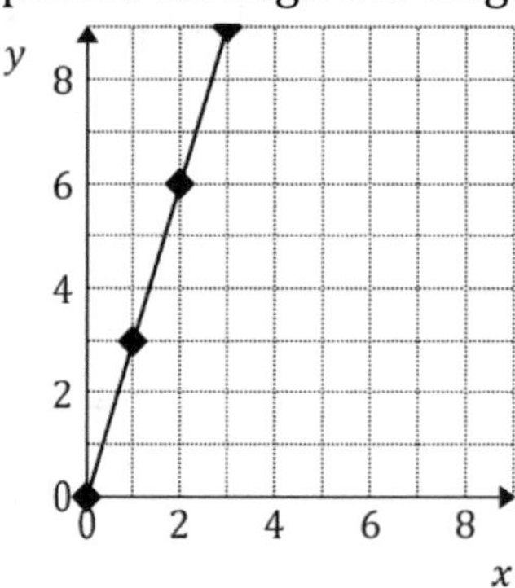

② Plot each (x, y) pair and then draw a line through the points. Each value of y is *one-half* as large as the corresponding value of x. The graph below has a rise of 4.5 and a run of 9. The slope is $\frac{4.5}{9} = \frac{1}{2}$. The y-intercept is 0: The line passes through the origin. The equation for this line is $y = \frac{x}{2}$.

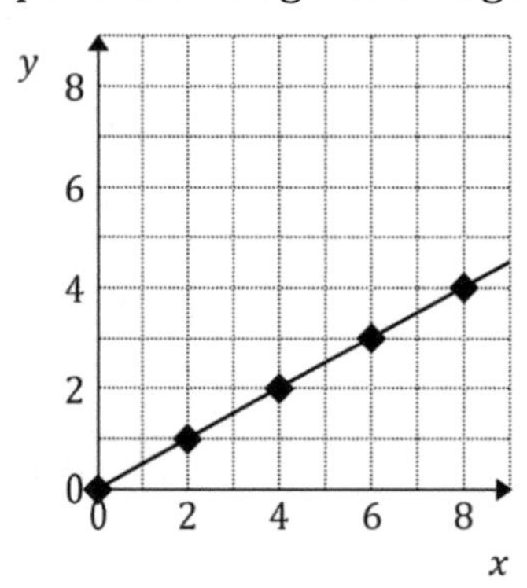

Page 246

① $2 \times 24 = \boxed{48}$ stripes and $2 \times 72 = \boxed{144}$ stripes

② $3 + 6 = \boxed{9}$ quarts and $9 + 6 = \boxed{15}$ quarts

Page 247

③ $16 \times 7 = \boxed{112}$ GB and $16 \times 25 = \boxed{400}$ GB

④ $\frac{3}{8} \times 40 = \frac{3 \times 40}{8} = \frac{120}{8} = \boxed{15}$ eggs and $\frac{3}{8} \times 64 = \frac{3 \times 64}{8} = \frac{192}{8} = \boxed{24}$ eggs

Page 248

⑤ $7 + 15 - 1 = 22 - 1 = \boxed{21}$ and $21 + 15 = \boxed{36}$

Exercise 5 is a "tricky" question. Starting on page 7, after reading 15 pages, the student will be ready to read page 22, meaning that the student will have finished reading page 21 (which is what the question asked). Let's try a simpler situation first. If you start on page 7 and read 2 pages, you will read page 7 and page 8, meaning that you will finish reading page 8, yet if you add 2 to 7 you get $7 + 2 = 9$. The problem is that we need to subtract 1 for page 7.

In this problem, the student reads pages 7, 8, 9, 10, 11, 12, 13, 14, 15, 16, 17, 18, 19, 20, and 21. If you count, you'll see that 21 is the 15th page read.

In the second part, the student reads pages 22 thru 36, where $22 + 15 - 1 = 36$.

⑥ $\frac{\$3}{5} \times 20 = \frac{\$3 \times 20}{5} = \frac{\$60}{5} = \boxed{\$12}$ and $\frac{\$3}{5} \times 30 = \frac{\$3 \times 30}{5} = \frac{\$90}{5} = \boxed{\$18}$

Page 249

⑦ $7 + 9 = \boxed{16}$ and $76 + 9 = \boxed{85}$

⑧ $32 \times 8 = \boxed{256}$ and $32 \times 29 = \boxed{928}$

Page 250

① (C) $96 + 17 = 113$

② (E) $648 \times 3 = 1944$

③ (B) multiplicative: $z = 2w$ (multiply the top row by 2 to get the bottom row)

④ (A) additive: $E = D + 7$ (add 7 to each number in the top row to get the bottom row) Note: Although 14 is twice 7, the other numbers on the bottom row aren't twice the corresponding number from the top row.

⑤ (A) $F = 6B$ (each number on the bottom row is 6 times the corresponding number from the top row)

⑥ (D) $h = d + 12$ (each number on the bottom row is 12 more than the corresponding number from the top row)

Page 251

⑦ (D) $45 + 28 = 73$

⑧ (E) $8 \times 13 = 104$

⑨ (A) additive

⑩ (B) multiplicative (z equals 0.3 times y)

⑪ (E) $91 = 7 \times 13$

⑫ (D) 73 is prime

Notes: $35 = 5 \times 7, 42 = 6 \times 7, 51 = 3 \times 17$, and $87 = 3 \times 29$

⑬ (B) 2, 41, and $2 + 41 = 43$ are all prime numbers

Notes: $2 + 19 = 21 = 3 \times 7, 3 + 17 = 20 = 4 \times 5, 3 + 29 = 32 = 4 \times 8$, and the problem with E is that $6 = 2 \times 3$ isn't prime (the only even number that is prime is 2)

⑭ (B) $1.7 + 0.425 = 2.125$ (which agrees with the cell below it: $2.125 \times 2 = 4.25$)

⑮ (B) $y = 2x$ (each value on the bottom row is twice the corresponding value from the top row)

Page 252

⑯ (D) $y = x + 3$ For each value of x, the corresponding value of y is 3 more. For example, when $x = 6, y = 9$. This is a purely additive graph because the y-intercept isn't zero (it is 3) and because the slope is 1.

⑰ (B) $y = 2x$ For each value of x, the corresponding value of y is twice the value of x. For example, when $x = 6, y = 12$. This is a purely multiplicative graph because the y-intercept is zero (it passes through the origin). Note that the slope is 2 (because the rise is twice the run). When the run is 6, the corresponding rise is 12.

⑱ (B) the graph *immediately* above is multiplicative (whereas the graph at the top of the page is additive); see the notes for questions 16 and 17 above

⑲ (B) $\$8 \times 14 = \$112$ (where the 14 comes from $42 \div 3 = 14$)

Chapter 8: Variables

Page 254

① y is the variable, 4 is the coefficient, and 4, 3, and 5 are constants. (If you wrote -3, perhaps you were thinking of adding the constant negative three rather than subtracting the constant three.)

② t is the variable, 2 and 5 are the coefficients, and 2, 9, and 5 are constants.

③ x is the variable, 6 is the coefficient, and 7, 6, and 25 are constants.

Page 256

Note: We are only giving the numerical answer and showing that it works. We didn't include the diagrams for these problems in the answer key.

① $x = 5$ because $5 + 4 = 9$

② $x = 7$ because $3 + 7 = 10$

③ $x = 9$ because $16 = 7 + 9$

④ $x = 8$ because $14 = 8 + 6$

⑤ $x = 6$ because $4 + 6 = 10$

⑥ $x = 1$ because $8 = 7 + 1$

⑦ $x = 9$ because $17 = 9 + 8$

⑧ $x = 3$ because $3 + 9 = 12$

Page 257

① $x = 12 - 6 = \boxed{6}$ Check: $6 + 6 = 12$

② $x = 8 - 3 = \boxed{5}$ Check: $3 + 5 = 8$

Page 258

③ $15 - 7 = \boxed{8} = x$ Check: $15 = 7 + 8$

④ $x = 13 - 9 = \boxed{4}$ Check: $4 + 9 = 13$

⑤ $x = 5 - 3 = \boxed{2}$ Check: $3 + 2 = 5$

⑥ $11 - 5 = \boxed{6} = x$ Check: $11 = 5 + 6$

⑦ $14 - 7 = \boxed{7} = x$ Check: $14 = 7 + 7$

⑧ $x = 12 - 7 = \boxed{5}$ Check: $5 + 7 = 12$

⑨ $x = 4 - 4 = \boxed{0}$ Check: $4 + 0 = 4$

⑩ $18 - 9 = \boxed{9} = x$ Check: $18 = 9 + 9$

⑪ $10 - 6 = \boxed{4} = x$ Check: $10 = 6 + 4$

⑫ $x = 16 - 8 = \boxed{8}$ Check: $8 + 8 = 16$

⑬ $x = 16 - 9 = \boxed{7}$ Check: $7 + 9 = 16$

⑭ $12 - 4 = \boxed{8} = x$ Check: $12 = 4 + 8$

Page 259

① $x = 6 + 3 = \boxed{9}$ Check: $9 - 3 = 6$

② $7 + 5 = \boxed{12} = x$ Check: $7 = 12 - 5$

Page 260

③ $4 + 4 = \boxed{8} = x$ Check: $4 = 8 - 4$

④ $x = 9 + 8 = \boxed{17}$ Check: $17 - 8 = 9$

⑤ $x = 5 + 6 = \boxed{11}$ Check: $11 - 6 = 5$

⑥ $8 + 5 = \boxed{13} = x$ Check: $8 = 13 - 5$

⑦ $7 + 8 = \boxed{15} = x$ Check: $7 = 15 - 8$

⑧ $x = 9 + 9 = \boxed{18}$ Check: $18 - 9 = 9$

⑨ $x = 0 + 6 = \boxed{6}$ Check: $6 - 6 = 0$

⑩ $3 + 5 = \boxed{8} = x$ Check: $3 = 8 - 5$

⑪ $5 + 9 = \boxed{14} = x$ Check: $5 = 14 - 9$

⑫ $x = 8 + 4 = \boxed{12}$ Check: $12 - 4 = 8$

⑬ $x = 1 + 1 = \boxed{2}$ Check: $2 - 1 = 1$

⑭ $6 + 8 = \boxed{14} = x$ Check: $6 = 14 - 8$

Page 261

① $x = 10 - 5 = \boxed{5}$ Check: $5 + 5 = 10$

② $7 + 9 = \boxed{16} = x$ Check: $7 = 16 - 9$

③ $x = 6 + 4 = \boxed{10}$ Check: $10 - 4 = 6$

④ $x = 14 - 8 = \boxed{6}$ Check: $6 + 8 = 14$

⑤ $7 + 6 = \boxed{13} = x$ Check: $7 = 13 - 6$

⑥ $11 - 7 = \boxed{4} = x$ Check: $11 = 7 + 4$

⑦ $13 - 5 = \boxed{8} = x$ Check: $13 = 5 + 8$

⑧ $x = 8 + 7 = \boxed{15}$ Check: $15 - 7 = 8$

⑨ $x = 16 - 7 = \boxed{9}$ Check: $9 + 7 = 16$

⑩ $9 + 2 = \boxed{11} = x$ Check: $9 = 11 - 2$

Page 264

Note: We are only giving the numerical answer and showing that it works. We didn't include the diagrams for these problems in the answer key.

① $x = 15$ because $15 + 30 = 45$

② $x = 72$ because $72 - 18 = 54$

③ $x = 14$ because $7 + 14 = 21$

④ $x = 32$ because $16 = 48 - 32$

⑤ $x = 36$ because $100 - 36 = 64$

⑥ $x = 1.7$ because $1.7 + 2.5 = 4.2$

⑦ $x = \frac{1}{2}$ because $\frac{1}{3} = \frac{1}{2} - \frac{1}{6}$ (because $\frac{1}{2} - \frac{1}{6} = \frac{3}{6} - \frac{1}{6} = \frac{2}{6} = \frac{1}{3}$)

⑧ $x = 50$ because $75 - 50 = 25$

Page 266

① $x = 24 \div 4 = \boxed{6}$ Check: $4 \times 6 = 24$

② $x = 14 \div 7 = \boxed{2}$ Check: $7 \times 2 = 14$

③ $15 \div 5 = \boxed{3} = x$ Check: $15 = 5 \times 3$

④ $x = 35 \div 5 = \boxed{7}$ Check: $5 \times 7 = 35$

⑤ $x = 12 \div 3 = \boxed{4}$ Check: $3 \times 4 = 12$

⑥ $48 \div 6 = \boxed{8} = x$ Check: $48 = 6 \times 8$

⑦ $81 \div 9 = \boxed{9} = x$ Check: $81 = 9 \times 9$

⑧ $x = 56 \div 8 = \boxed{7}$ Check: $8 \times 7 = 56$

⑨ $x = 5 \div 5 = \boxed{1}$ Check: $5 \times 1 = 5$

⑩ $63 \div 7 = \boxed{9} = x$ Check: $63 = 7 \times 9$

Page 268

① $x = 4 \times 7 = \boxed{28}$ Check: $\frac{28}{4} = 28 \div 4 = 7$

② $x = 8 \times 4 = \boxed{32}$ Check: $\frac{32}{8} = 32 \div 8 = 4$

③ $9 \times 2 = \boxed{18} = x$ Check: $9 = \frac{18}{2} = 18 \div 2$

④ $x = 5 \times 5 = \boxed{25}$ Check: $\frac{25}{5} = 25 \div 5 = 5$

⑤ $x = 7 \times 6 = \boxed{42}$ Check: $\frac{42}{7} = 42 \div 7 = 6$

⑥ $x = 9 \times 3 = \boxed{27}$ Check: $\frac{27}{9} = 27 \div 9 = 3$

⑦ $4 \times 3 = \boxed{12} = x$ Check: $4 = \frac{12}{3} = 12 \div 3$

⑧ $3 \times 6 = \boxed{18} = x$ Check: $3 = \frac{18}{6} = 18 \div 6$

⑨ $x = 9 \times 6 = \boxed{54}$ Check: $\frac{54}{9} = 54 \div 9 = 6$

⑩ $9 \times 8 = \boxed{72} = x$ Check: $9 = \frac{72}{8} = 72 \div 8$

Page 270

① $x = 27 + 48 = \boxed{75}$ Check: $75 - 27 = 48$

② $x = \frac{148}{4} = 148 \div 4 = \boxed{37}$ (see Chapter 2) Check: $4 \times 37 = 148$

Note: For most of these problems, to do the arithmetic, you should have a longer solution that applies a method from Chapters 1-5 (such as how to multiply or divide with multi-digit numbers).

③ $x = 17 \times 26 = \boxed{442}$ (see Chapter 1) Check: $\frac{442}{17} = 442 \div 17 = 26$

④ $8.3 - 3.5 = \boxed{4.8} = x$ (see Chapter 4) Check: $8.3 = 4.8 + 3.5$

⑤ $\frac{156}{13} = 156 \div 13 = \boxed{12} = x$ (see Chapter 2) Check: $13 \times 12 = 156$

⑥ $\frac{2}{3} + \frac{1}{12} = \frac{8}{12} + \frac{1}{12} = \frac{9}{12} = \boxed{\frac{3}{4}} = x$ (see Chapter 5) Check: $\frac{2}{3} = \frac{3}{4} - \frac{1}{12}$ because $\frac{3}{4} - \frac{1}{12} = \frac{9}{12} - \frac{1}{12} = \frac{8}{12} = \frac{2}{3}$

⑦ $x = \frac{7}{4} \div 9 = \frac{7}{4 \times 9} = \boxed{\frac{7}{36}}$ (see Chapter 5) Check: $9 \times \frac{7}{36} = \frac{9 \times 7}{36} = \frac{63}{36} = \frac{63 \div 9}{36 \div 9} = \frac{7}{4}$

⑧ $x = 7 \times \frac{3}{14} = \frac{7 \times 3}{14} = \frac{21}{14} = \boxed{\frac{3}{2}}$ (see Chapter 5) Check: $\frac{3}{2} \div 7 = \frac{3}{2 \times 7} = \frac{3}{14}$

⑨ $x = 431 - 276 = \boxed{155}$ Check: $155 + 276 = 431$

⑩ $\frac{3.2}{4} = 3.2 \div 4 = \boxed{0.8} = x$ (see Chapter 4) Check: $3.2 = 4 \times 0.8$

Page 272

① Step 1: $3x = 18$ Step 2: $x = \boxed{6}$ Check: $3 \times 6 + 7 = 18 + 7 = 25$

② Step 1: $8x = 40$ Step 2: $x = \boxed{5}$ Check: $8 \times 5 - 18 = 40 - 18 = 22$

③ Step 1: $\frac{x}{3} = 9$ Step 2: $x = \boxed{27}$ Check: $\frac{27}{3} - 4 = 9 - 4 = 5$

④ Step 1: $36 = 6x$ Step 2: $\boxed{6} = x$ Check: $60 = 6 \times 6 + 24 = 36 + 24$

Page 273

⑤ Step 1: $\frac{x}{7} = 7$ Step 2: $x = \boxed{49}$ Check: $\frac{49}{7} + 2 = 7 + 2 = 9$

⑥ Step 1: $36 = 4x$ Step 2: $\boxed{9} = x$ Check: $21 = 4 \times 9 - 15 = 36 - 15$

⑦ Step 1: $42 = 7x$ Step 2: $\boxed{6} = x$ Check: $61 = 19 + 7 \times 6 = 19 + 42$

Note: In the check above, recall from Sec. 1.12 that we multiply before we add.

⑧ Step 1: $4 = \frac{x}{6}$ Step 2: $\boxed{24} = x$ Check: $2 = \frac{24}{6} - 2 = 4 - 2$

⑨ Step 1: $\frac{x}{8} = 8$ Step 2: $x = \boxed{64}$ Check: $\frac{64}{8} - 1 = 8 - 1 = 7$

⑩ Step 1: $72 = 9x$ Step 2: $\boxed{8} = x$ Check: $100 = 28 + 9 \times 8 = 28 + 72$

Note: In the check above, recall from Sec. 1.12 that we multiply before we add.

Page 275

① Model: $x - 18 = 24$ Answer: $x = \$42$

② Model: $6x = 102$ Answer: $x = 17$ trading cards

Note: These are not the only equations that can be used to model these problems.

Page 276

③ Model: $7 + x = 11$ Answer: $x = 4$ hats

④ Model: $\frac{x}{9} = 13$ Answer: $x = 117$ marbles

Page 277

⑤ Model: $x + 8 = 27$ Answer: $x = 19$ pictures

⑥ Model: $6x = 192$ Answer: $x = 32$ inches

Page 278

⑦ Model: $\frac{x}{12} = 14$ Answer: $x = 168$ seeds

⑧ Model: $x - 7 = 17$ Answer: $x = 24$ flowers

Page 279

① (C) $x = 8 - 3 = \boxed{5}$ (starting from $x + 3 = 8$)

② (C) $x = 50 - 25 = \boxed{25}$ Check: $25 + 25 = 50$

③ (D) $x = 48 + 16 = \boxed{64}$ Check: $64 - 16 = 48$

④ (B) $x = \frac{72}{9} = 72 \div 9 = \boxed{8}$ Check: $9 \times 8 = 72$

⑤ (E) $x = 4 \times 24 = \boxed{96}$ Check: $\frac{96}{4} = 96 \div 4 = 24$

⑥ (B) $\boxed{x + 12 = 20}$ The two bottom strips add up to the same length as the top strip. Note that $x = 8$.

Page 280

⑦ (C) $x = \boxed{14}$ Step 1: $6x = 84$ Step 2: $x = \frac{84}{6} = 84 \div 6 = 14$

Check: $6x - 30 = 6 \times 14 - 30 = 84 - 30 = 54$

⑧ (D) $x = \boxed{32}$ Step 1: $\frac{x}{4} = 8$ Step 2: $x = 4 \times 8 = 32$

Check: $\frac{x}{4} + 12 = \frac{32}{4} + 12 = 32 \div 4 + 12 = 8 + 12 = 20$

⑨ (B) $\boxed{x - 18 = 36}$

⑩ (C) $x = 36 + 18 = \boxed{54}$ Check: $54 - 18 = 36$

⑪ (D) $\boxed{\frac{x}{8} = 24}$

⑫ (E) $x = 8 \times 24 = \boxed{192}$ Check: $\frac{192}{8} = 192 \div 8 = 24$

Chapter 9: Measurement and Geometry

Page 285

① A yardstick (a child's tennis racket is longer than a foot, but less than a yard)

② a ruler (the width of a piece of notebook paper is shorter than a foot)

③ a flexible tape measure (because a flexible tape measure can wrap around a person's waist, whereas a yardstick is usually rigid)

④ a tape measure (a doorway is taller than a yard; a yardstick could be used, but you would have to mark spots carefully and add distances together)

⑤ a trundle wheel (you can walk along the length of a basketball court)

⑥ a car's odometer (a person can usually drive between two cities)

⑦ a balance (to measure the weight of a small item)

⑧ a suitable measuring cup (to measure capacity)

Page 286

⑨ a combination of feet and inches is common (like 5 feet, 8 inches)

⑩ pounds (a typical bathroom scale gives weight in pounds; though a doctor's office may use a combination of pounds and ounces)

⑪ gallons (most gas stations in the United States sell gas by the gallon)

⑫ fluid ounces (medicine usually comes in very small doses)

⑬ miles (since a car's odometer measures miles)

⑭ inches (are common when shopping for a suit or a shirt)

Page 288

① $1 \text{ cg} = \frac{1}{100} \text{ g}$ or 0.01 g

② $1 \text{ mL} = \frac{1}{1000} \text{ L}$ or 0.001 L

③ $1 \text{ km} = 1000 \text{ m}$

④ $1 \text{ mg} = \frac{1}{1000} \text{ g}$ or 0.001 g

⑤ $1 \text{ cL} = \frac{1}{100} \text{ L}$ or 0.01 L

⑥ $1 \text{ dkg} = 10 \text{ g}$

⑦ $1 \text{ dL} = \frac{1}{10} \text{ L}$ or 0.1 L

⑧ $1 \text{ hm} = 100 \text{ m}$

Page 292

① $36 \text{ ft} \times \frac{12 \text{ in.}}{1 \text{ ft}} = 36 \times 12 \text{ in.} = \boxed{432 \text{ in.}}$ Note: Going from feet to inches, we multiply because an inch (the final unit) is smaller than a foot (the initial unit).

② $21 \text{ ft} \times \frac{1 \text{ yd}}{3 \text{ ft}} = 21 \div 3 \text{ yd} = \boxed{7 \text{ yd}}$ Note: Going from feet to yards, we divide because a yard (the final unit) is larger than a foot (the initial unit).

③ $15 \text{ yd} \times \frac{3 \text{ ft}}{1 \text{ yd}} = 15 \times 3 \text{ ft} = \boxed{45 \text{ ft}}$ Note: Going from yards to feet, we multiply because a foot (the final unit) is smaller than a yard (the initial unit).

④ $168 \text{ in.} \times \frac{1 \text{ ft}}{12 \text{ in.}} = 168 \div 12 \text{ ft} = \boxed{14 \text{ ft}}$ Note: Going from inches to feet, we divide because a foot (the final unit) is larger than an inch (the initial unit).

⑤ Step 1: $4 \text{ yd} \times \frac{3 \text{ ft}}{1 \text{ yd}} = 4 \times 3 = 12 \text{ ft}$ (not finished yet)

Step 2: $12 \text{ ft} \times \frac{12 \text{ in.}}{1 \text{ ft}} = 12 \times 12 \text{ in.} = \boxed{144 \text{ in.}}$ Note: Going from yards to inches, we multiply because an inch (the final unit) is smaller than a yard (the initial unit).
Note: An alternate solution is to use 1 yd = 36 in. in a single step.

Page 293

⑥ $32 \text{ pt} \times \frac{16 \text{ fl oz}}{1 \text{ pt}} = 32 \times 16 \text{ fl oz} = \boxed{512 \text{ fl oz}}$ Note: Going from pints to fluid ounces, we multiply because a fluid ounce (the final unit) is smaller than a pint (the initial unit).
Note: If you check these answers with an online conversion calculator, be sure to select "**US liquid pint**" and **not** "pint" (or you'll get a different answer).

⑦ $48 \text{ qt} \times \frac{1 \text{ gal}}{4 \text{ qt}} = 48 \div 4 \text{ gal} = \boxed{12 \text{ gal}}$ Note: Going from quarts to gallons, we divide because a gallon (the final unit) is larger than a quart (the initial unit).

⑧ $16 \text{ gal} \times \frac{8 \text{ pt}}{1 \text{ gal}} = 16 \times 8 \text{ pt} = \boxed{128 \text{ pt}}$ Note: Going from gallons to pints, we multiply because a pint (the final unit) is smaller than a gallon (the initial unit).

⑨ $48 \text{ lb} \times \frac{16 \text{ oz}}{1 \text{ lb}} = 48 \times 16 \text{ oz} = \boxed{768 \text{ oz}}$ Note: Going from pounds to ounces, we multiply because an ounce (the final unit) is smaller than a pound (the initial unit).

⑩ $80 \text{ oz} \times \frac{1 \text{ lb}}{16 \text{ oz}} = 80 \div 16 \text{ lb} = \boxed{5 \text{ lb}}$ Note: Going from ounces to pounds, we divide because a pound (the final unit) is larger than an ounce (the initial unit).

Page 295

① 2.7 kg = 2700 g (from k to "none" is 3 steps to the right, so we multiply by 1000, which shifts the decimal point 3 places to the right)

Page 296

② 54 mL = 0.054 L (going from m to "none" is 3 steps to the left, so we divide by 1000, which shifts the decimal point 3 places to the left)

③ 3.15 cm = 0.0315 m (going from c to "none" is 2 steps to the left, so we divide by 100, which shifts the decimal point 2 places to the left)

④ 0.42 dg = 0.0042 dkg (going from d to dk is 2 steps to the left, so we divide by 100, which shifts the decimal point 2 places to the left)

⑤ 4 hL = 40,000 cL (from h to c is 4 steps to the right, so we multiply by 10,000, which shifts the decimal point 4 places to the right)

⑥ 9.6 kg = 9,600,000 mg (from k to m is 6 steps to the right, so we multiply by 1,000,000, which shifts the decimal point 6 places to the right)

Page 298

① 40 in. < 5 ft because $5 \text{ ft} \times \frac{12 \text{ in.}}{1 \text{ ft}} = 5 \times 12 \text{ in.} = 60 \text{ in.}$

② 0.4 L > 300 mL because $0.4 \text{ L} = 0.4 \times 1000 \text{ mL} = 400 \text{ mL}$ (going from "none" to m is 3 steps to the right, so we multiply by 1000, which shifts the decimal point 3 places to the right)

③ 2.4 kg > 2000 g because $2.4 \text{ kg} = 2.4 \times 1000 \text{ g} = 2400 \text{ g}$ (going from kg to "none" is 3 steps to the right, so we multiply by 1000, which shifts the decimal point 3 places to the right)

④ 8 qt = 2 gal because $2 \text{ gal} \times \frac{4 \text{ qt}}{1 \text{ gal}} = 2 \times 4 \text{ qt} = 8 \text{ qt}$

⑤ 3.5 T < 7500 lb because $3.5 \text{ T} \times \frac{2000 \text{ lb}}{1 \text{ T}} = 3.5 \times 2000 \text{ lb} = 7000 \text{ lb}$

Note: This is a "**United States ton**," not a metric ton.

⑥ 12 hm > 0.9 km because $0.9 \text{ km} = 0.9 \times 10 \text{ hm} = 9 \text{ hm}$ (going from k to h is 1 step to the right, so we multiply by 10, which shifts the decimal point 1 place to the right)

⑦ 7 lb > 100 oz because $7 \text{ lb} \times \frac{16 \text{ oz}}{1 \text{ lb}} = 7 \times 16 \text{ oz} = 112 \text{ oz}$

⑧ 7 dg < 0.8 g because 0.8 g = 0.8 × 10 dg = 8 dg (going from "none" to d is 1 step to the right, so we multiply by 10, which shifts the decimal point 1 place to the right)

⑨ 1.4 cL = 14 mL because 1.4 cL = 1.4 × 10 mL = 14 mL (going from c to m is 1 step to the right, so we multiply by 10, which shifts the decimal point 1 place to the right)

⑩ 75 in. < 3 yd because $3 \text{ yd} \times \frac{36 \text{ in.}}{1 \text{ yd}} = 3 \times 36 \text{ in.} = 108 \text{ in.}$

Note: 1 yd = 36 in. because 1 yd = 3 ft and 1 ft = 12 in. (since 3 × 12 = 36)

Page 299

① Step 1: 40 × 8 oz = 320 oz (not finished yet)

Step 2: $320 \text{ oz} \times \frac{1 \text{ lb}}{16 \text{ oz}} = 320 \div 16 \text{ lb} = \boxed{20 \text{ lb}}$

Page 300

② Step 1: 0.6 km = 0.6 × 1000 m = 600 m (not finished yet)

Step 2: 600 m ÷ 50 m = $\boxed{12}$ sprints

③ Step 1: 12 × 500 lb = 6000 lb (not finished yet)

Step 2: $6000 \text{ lb} \times \frac{1 \text{ T}}{2000 \text{ lb}} = 6000 \div 2000 \text{ T} = \boxed{3 \text{ T}}$

Note: This is a "**United States ton**," not a metric ton.

④ Step 1: $2 \text{ gal} \times \frac{128 \text{ fl oz}}{1 \text{ gal}} = 2 \times 128 \text{ fl oz} = 256 \text{ fl oz}$ (not finished yet)

Step 2: 256 fl oz ÷ 8 = $\boxed{32 \text{ fl oz}}$

Note: 1 gal = 128 fl oz because 1 gal = 4 qt, 1 qt = 2 pt, and 1 pt = 16 fl oz (since 4 × 2 × 16 = 8 × 16 = 128)

⑤ Step 1: $6 \text{ ft} \times \frac{12 \text{ in.}}{1 \text{ ft}} = 6 \times 12 \text{ in.} = 72 \text{ in.}$ (not finished yet)

Step 2: 72 in. − 55 in. = $\boxed{17 \text{ in.}}$

Page 302

① both angles are right

② obtuse angle (on the left) and acute angle (on the right)

③ 70° is an acute angle and 130° is an obtuse angle

④ 90° is a right angle and 75° is an acute angle

Page 304

① regular pentagon (5 sides, equilateral and equiangular)

② irregular quadrilateral (4 sides)

③ irregular octagon (8 sides, equilateral but not equiangular because one-half of the angles are larger than 180°, meaning that they bend outward, whereas the other half of the angles are smaller than 180°)

Note: Although we didn't discuss the terms convex or concave, star-shaped polygons are concave, whereas the polygons in Exercises 1, 2, 4, 6, 7, and 8 are convex. A concave polygon has an interior angle that is greater than 180°. Also, every diagonal of a convex polygon lies inside of the polygon, whereas a concave polygon has at least one diagonal that lies outside of the polygon. A concave polygon doesn't meet our definition of regular because a concave polygon can't be equiangular.

④ regular heptagon (7 sides, equilateral and equiangular)

⑤ irregular hexagon (6 sides, equilateral but not equiangular because one of the angles is greater than 180°)

Note: See the note to the solution to Exercise 3. This polygon is also concave because one diagonal lies outside of the polygon.

⑥ irregular triangle (3 sides)

Note: In Sec. 9.9, we'll learn that this triangle is isosceles because 2 of its sides are congruent. (Note that one side is clearly longer than the other 2.)

⑦ irregular decagon (10 sides, equiangular but not equilateral because the top and bottom sides are clearly longer than the other 8)

⑧ irregular nonagon (9 sides, clearly not equiangular because 2 of the angles are 126° whereas the other 7 angles are 144°)

Page 306

① right scalene

② acute isosceles

③ acute, equilateral, and also isosceles (because isosceles means at least two sides are congruent; since all three are congruent, this satisfies the definition)

④ obtuse scalene

Page 308

① square (four 90° angles and four congruent sides)

② trapezoid (one pair of parallel edges)

③ parallelogram (two pairs of parallel edges)

④ rectangle (four 90° angles)

⑤ rhombus (four congruent sides)

⑥ quadrilateral is the best word defined in this section (if you know the term "kite," that is more precise; the kite **isn't** a parallelogram or trapezoid)

Page 310

① $P = 4L = 4 \times 12 = \boxed{48}$ and $A = L^2 = 12 \times 12 = \boxed{144}$

② $P = 2L + 2W = 2 \times 20 + 2 \times 16 = 40 + 32 = \boxed{72}$ and $A = LW = 20 \times 16 = \boxed{320}$

③ $P = 20 + 16 + 12 = \boxed{48}$ and $A = \frac{1}{2}bh = \frac{1}{2} \times 16 \times 12 = 8 \times 12 = \boxed{96}$

④ $P = 18 + 18 + 11 + (18 - 11) + 6 + (18 - 6) = 47 + 7 + 6 + 12 = \boxed{72}$ and $A = 18 \times 18 - (18 - 11) \times (18 - 6) = 324 - 7 \times 12 = 324 - 84 = \boxed{240}$

Note: This polygon can be formed by removing a 7 by 12 rectangle, as illustrated below. The length of the removed rectangle is $18 - 6 = 12$ and the width of the removed rectangle is $18 - 11 = 7$. The area of the removed rectangle is $12 \times 7 = 84$. Subtract 84 from the area of the square from which it was removed to get the area of the polygon: $18^2 - 84 = 324 - 84 = 240$. To find the perimeter, add all six sides: $18 + 18 + 11 + 12 + 7 + 6 = 36 + 23 + 13 = 72$.

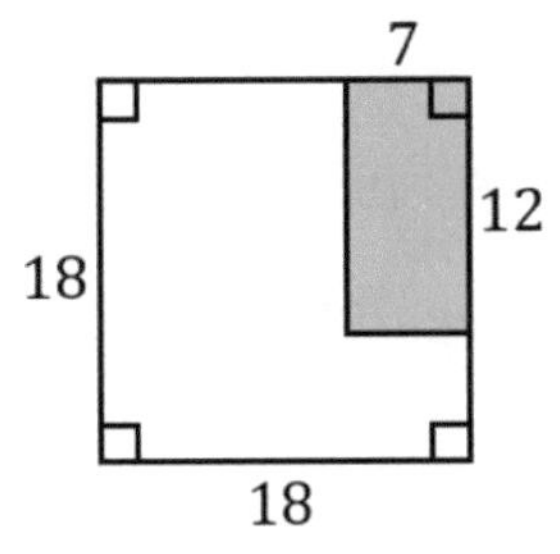

Page 312

① $V = LWH = 20 \times 8 \times 9 = 20 \times 72 = 1440$

② $V = L^3 = 7 \times 7 \times 7 = 49 \times 7 = 343$

③ $V = LWH = 12 \times 10 \times 4 = 120 \times 4 = 480$

④ $V = L^3 = 8 \times 8 \times 8 = 64 \times 8 = 512$

⑤ $V = LWH = 6 \times 6 \times 5 = 36 \times 5 = 180$

Alternate method: $A = L^2 = 6 \times 6 = 36$ and $V = AH = 36 \times 5 = 180$

Page 314

① $V = LWH = 8 \times 6 \times 3 = 48 \times 3 = 144$

② $V = L^3 = 5 \times 5 \times 5 = 25 \times 5 = 125$

③ $V = LWH = 12 \times 5 \times 2 = 60 \times 2 = 120$

④ $V = LWH = 10 \times 5 \times 5 = 50 \times 5 = 250$

Page 315

① (D) trundle wheel (you could walk along the track)

② (A) balance (to measure mass; the mass is way too small for a truck scale)

③ (C) 26 yd Notes: In miles, the answer would be 0.015 mi, which is 10 times smaller than A. In feet, the answer would be 78 ft, which is much bigger than B. Since a meter is a few inches longer than a yard, 78 m is way too big. (78 ft would have been correct, but that isn't one of the answers.) In inches, the answer would be 936 in., which is about 10 times greater than E.

④ (E) 12 fl oz equates to 1.5 c, which is reasonable for a coffee mug. Notes: 2 gal is way too big for a mug. 6 qt is 1.5 gal, again much too large. 8 pt is 4 qt which is 1 gal, which is again too large for a mug. 10 c is enough for 10 mugs, but is too much for 1 mug.

⑤ (B) 1 cL = 0.03 L because centi stands for $\frac{1}{100} = 0.01$

⑥ (B) $24 \text{ ft} \times \frac{1 \text{ yd}}{3 \text{ ft}} = 24 \div 3 \text{ yd} = \boxed{8 \text{ yd}}$

⑦ (E) $48 \text{ qt} \times \frac{4 \text{ c}}{1 \text{ qt}} = 48 \times 4 \text{ pt} = \boxed{192 \text{ c}}$ Note: 1 qt = 4 c because 1 qt = 2 pt and 1 pt = 2 c (and since $2 \times 2 = 4$) Note: We are using **United States pints**.

⑧ (A) $272 \text{ oz} \times \frac{1 \text{ lb}}{16 \text{ oz}} = 272 \div 16 \text{ lb} = \boxed{17 \text{ lb}}$

Page 316

⑨ (B) 37.4 cm = 0.374 m (going from c to "none" is 2 steps to the left, so we divide by 100, which shifts the decimal point 2 places to the left)

Alternate solution: centi stands for $\frac{1}{100}$ and $37.4 \div 100 = 0.374$

⑩ (E) 6.2 kg = 6200 g (from k to "none" is 3 steps to the right, so we multiply by 1000, which shifts the decimal point 3 places to the right)
Alternate solution: kilo stands for 1000 and $6.2 \times 1000 = 6200$
⑪ (B) 5 yd Notes: 5 yd = 15 ft (which is greater than C) 5 yd = 180 in. (which is greater than E) 5 yd > 2 m (because a meter is only slightly larger than a yard) 5 yd > 87 cm because 87 cm is smaller than a meter (which is smaller than A)
⑫ (E) is the incorrect inequality because 1 qt = 36 c, which isn't less than 30 c
Notes: 1 qt = 4 c because 1 qt = 2 pt and 1 pt = 2 c (and since $2 \times 2 = 4$)
3 c = 24 fl oz (which is greater than 20 fl oz, so A is true)
2 gal = 8 qt (which is less than 9 qt, so B is true)
5 gal = 40 pt (which is greater than 35 pt, so C is true; 1 gal = 8 pt because 1 gal = 4 qt and 1 qt = 2 pt; note that $4 \times 2 = 8$)
7 pt = 14 c (which is less than 15 c, so D is true)
Note: If you check these answers with an online conversion calculator, be sure to select "**US liquid pint**" and **not** "pint" (or you'll get a different answer).
⑬ (A) 24 ft because 1 yd = 3 ft and 12 in. = 1 ft, such that
3×2 yd + 4×18 in. = 6 yd + 72 in. = 18 ft + 6 ft = 24 ft
Alternate solution: 1 yd = 3 ft, so one 2 yd chain is 6 ft long, such that 3 of these chains makes 18 feet; 12 in. = 1 ft, so one 18 in. chain is 1.5 ft, such that 4 of these chains makes 6 feet; 18 ft + 6 ft = 24 ft
Notes: 24 ft equates to 288 inches; D and E are incorrect because they have the wrong units with the number 288
B is unreasonably small, whereas C is unreasonably large
⑭ (D) octagon (a polygon with 8 sides)
⑮ (C) regular (which means both equilateral **AND** equiangular; A and B are incorrect because of the word "only")

Page 317

⑯ (C) octagon (a polygon with 8 sides)
Note: 14 and 16 are both octagons because they both have 8 sides, even though the pictures look much different.

⑰ (B) equilateral only (it isn't equiangular because the shape is concave, meaning that at least one interior angle is greater than 180°, as discussed in the solutions to Exercises 3 and 5 in Sec. 9.8; see page 414)
Note that C is incorrect because the polygon isn't equiangular. A polygon needs to be BOTH equilateral AND equiangular in order to be regular.
⑱ (B) right, isosceles (it is right because it has a 90° angle and it is isosceles because two sides are congruent)
⑲ (B) $17 + 17 + 24 = 34 + 24 = 58$ (add up all three sides)
⑳ (B) $A = \frac{1}{2}bh = \frac{1}{2} \times 17 \times 17 = \frac{1}{2} \times 289 = \frac{289}{2} = 189 \div 2 = 144.5$
Note that the base is 17 and the height is also 17. (The hypotenuse, 24, is not needed for this calculation.)
Alternate solution: Draw a second triangle congruent to the first triangle to make a square with sides of 17. The area of the square is $17 \times 17 = 289$. The triangle has half the area of the square, which is $\frac{289}{2} = 144.5$.
㉑ (C) $A = \frac{1}{2}bh = \frac{1}{2} \times 9 \times 6 = \frac{1}{2} \times 54 = \frac{54}{2} = 54 \div 2 = 27$

Page 318

㉒ (B) rectangle (four 90° angles, two pairs of congruent edges)
㉓ (D) $P = 2L + 2W = 2 \times 15 + 2 \times 9 = 30 + 18 = 48$
Alternate solution: $P = 15 + 9 + 15 + 9 = 24 + 24 = 48$
㉔ (D) $A = LW = 15 \times 9 = 135$
㉕ (C) $10 \times 8 - 7 \times 3 = 80 - 21 = 59$ (as shown below, subtract the area of the gray rectangle, $7 \times 3 = 21$, from the area of the large rectangle, $10 \times 8 = 80$, to get $80 - 21 = 59$)

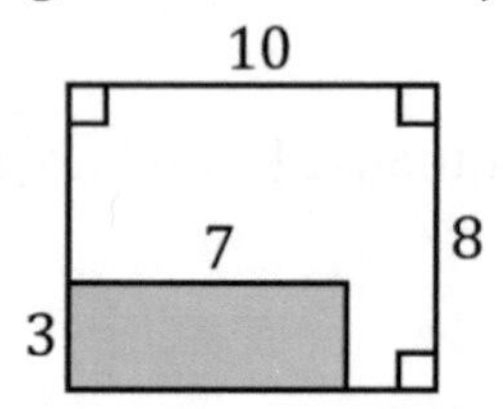

㉖ (E) $V = L^3 = 4 \times 4 \times 4 = 16 \times 4 = 64$
㉗ (D) $A = LW = 7 \times 4 = 28$
㉘ (E) $V = LWH = 7 \times 6 \times 5 = 42 \times 5 = 210$

Chapter 10: Finance

Page 320

① $37 \times \$0.25 = \boxed{\$9.25}$

Note: You should show more work in your solutions to these exercise, like the math we did in Chapter 4.

② $63 \times \$0.05 = \boxed{\$3.15}$

Page 321

③ $28 \times \$0.05 + 34 \times \$0.01 = \$1.40 + \$0.34 = \boxed{\$1.74}$

④ $17 \times \$0.10 + 35 \times \$0.05 = \$1.70 + \$1.75 = \boxed{\$3.45}$

⑤ $14 \times \$0.25 + 9 \times \$0.10 = \$3.50 + \$0.90 = \boxed{\$4.40}$

⑥ $73 \times \$0.25 + 42 \times \$0.10 + 29 \times \$0.05 = \$18.25 + \$4.20 + \$1.45 = \boxed{\$23.90}$

⑦ $22 \times \$0.25 + 11 \times \$0.10 + 17 \times \$0.05 + 56 \times \$0.01$

$= \$5.50 + \$1.10 + \$0.85 + \$0.56 = \boxed{\$8.01}$

⑧ $45 \times \$0.25 + 24 \times \$0.10 + 18 \times \$0.05 + 78 \times \$0.01$

$= \$11.25 + \$2.40 + \$0.90 + \$0.78 = \boxed{\$15.33}$

Page 323

① $\$15.19 - \$14.00 = \$1.19$ This is a sales tax. It is paid just once at the time of purchase (not annually). However, if the customer buys more books (or any other items) in the future, the customer will pay sales tax for those items, too.

② $\$18{,}500 - \$18{,}000 = \$500$ This is a property tax. It is typically an annual tax. Note that the amount of the property tax may change over time because the value of the land may change and the rate of the property tax may change.

Page 324

③ $3 \times \$0.08 = \boxed{\$0.24}$ is the total tax. This is a sales tax.

$\$3.00 + \$0.24 = \boxed{\$3.24}$ is the total amount paid.

④ \$80,000 has 80 thousands. \$25 is paid for each thousand dollars.

$\$25 \times 80 = \boxed{\$2000}$ is the annual tax. This is a property tax.

⑤ \$150 has 15 tens. \$1 is paid for each ten dollars.

$15 \times \$1 = \boxed{\$15}$ is the total tax. This is a sales tax.

$\$150 + \$15 = \boxed{\$165}$ is the total amount paid.

Page 326

① \$1800 has 18 hundreds. \$6 is withheld for each hundred dollars earned.

$18 \times \$6 = \boxed{\$108}$ is the total tax withheld each month. This is a payroll tax.

② \$42,000 has 420 hundreds (since $420 \times \$100 = \$42{,}000$). For the federal government, \$7 is withheld for each hundred dollars earned: $420 \times \$7 = \$2940$ is the federal tax. For the state, \$2 is withheld for each hundred dollars earned: $420 \times \$2 = \$840$ is the state tax. The total amount paid in taxes is $\$2940 + \$840 = \boxed{\$3780}$. These are federal and state **income** taxes.

Page 327

① $\$360 - \$75 = \boxed{\$285}$ per week

Page 328

② $\$1540 + \$135 + \$28 = \boxed{\$1703}$ per month

Note: Compare Problems 1 and 2. In Problem 1, we were given the gross and income and solved for the net income (so we subtracted). In Problem 2, we were given the net income and solved for the gross income (so we added).

③ $\$640 - \$48 - \$578 = \boxed{\$14}$

Check: The total taxes are $\$48 + \$14 = \$62$, which agrees with $\$640 - \$62 = \$578$.

④ $\$1250 - \$83 - \$26 - \$9 = \boxed{\$1132}$

Page 330

① Step 1: total income = $\$40 + \$32 = \$72$ (not the final answer)

Step 2: partial expenses = $\$12 + \$24 + \$17 = \$53$ (not the final answer)

Step 3: savings = total income – partial expenses = $\$72 - \$53 = \boxed{\$19}$

② Step 1: partial income = $\$50 + \$25 = \$75$ (not the final answer)

Step 2: total expenses = $\$18 + \$22 + \$55 + \$25 = \$120$ (not the final answer)

Step 3: tutoring income = total expenses – partial income = $\$120 - \$75 = \boxed{\$45}$

Second question: $\$150 \div \$25 = \boxed{6}$ months

Page 331

① debit (the withdrawal decreases your account balance)

② credit (the deposit increases your account balance)

③ debit (the purchase decreases your account balance)

④ debit (paying by check decreases your account balance)

Page 333

① Find the answers in the record below.

Tip: Check your final answer by adding up all of the debits and credits.

The total of the credits is $\$25 + \$35 + \$25 = \$85$.

The total of the debits is $\$3 + \$9 + \$16 + \$4 + \$23 + \$18 = \$73$.

Since the credits are $\$85 - \$73 = \$12$ more than the debits, the final balance should be $8 more than the initial balance. The final balance of $243 is $12 more than the beginning balance of $231.

Date	Description	Credit	Debit	Balance
	Beginning balance			$231
9/2	Bus fare		$3	$228
9/3	Allowance	$25		$253
9/5	Movie		$9	$244
9/5	Bookstore		$16	$228
9/7	Tutoring	$35		$263
9/8	Club fees		$4	$259
9/10	Allowance	$25		$284
9/10	Supplies		$23	$261
9/12	Miscellaneous		$18	$243

Page 334

① (D) $41 \times \$0.25 = \$10.25$

Note: A roll of quarters has 40 quarters valued at $10. This question has one more quarter than a roll of quarters has.

② (A) $67 \times \$0.05 = \$3.35$

Alternate method: $67 \times 5 = 335$ cents and 335 cents = $3.35

③ (B) $\$8.75 \div \$0.25 = 35$ quarters Check: $35 \times \$0.25 = \$8.75$

Alternate method: Convert to cents before dividing: $875 \div 25 = 35$

④ (C) $11 \times \$0.25 + 3 \times \$0.10 + 7 \times \$0.05 + 18 \times \$0.01$
$= \$2.75 + \$0.30 + \$0.35 + \$0.18 = \$3.58$

Alternate method: Work with cents first and then divide by 100.

⑤ (D) sales tax

⑥ (A) income tax

⑦ (C) property tax

Page 335

⑧ (C) $\$31.86 - \$29.50 = \$2.36$

⑨ (D) $7 \times \$0.05 = \$0.35$

⑩ (D) $90 \times \$45 = \$4050$

⑪ (A) $\$1400 - \$79 - \$23 = \$1298$

⑫ (C) payroll deposit (adds money to your bank account)

⑬ (A) ATM withdrawal (subtracts money from your bank account)

Page 336

⑭ (B) Income exceeds expenses. The total income is \$115, whereas the total expenses are \$105.

⑮ (B) Add \$10 to savings. This will make the total expenses \$115, matching the total income.

⑯ (A) $\$80 \div \$16 = 5$ months

⑰ (A) $\$327 - \$58 = \$269$

GLOSSARY

acute: an angle with an angular measure that is less than 90°. An acute triangle has three acute angles.

associative property of multiplication: when multiplying three numbers, the grouping of the numbers doesn't affect the answer: $(a \times b) \times c = a \times (b \times c)$.

bar graph: a graph with vertical bars that depict frequency.

base: the bottom side of a triangle.

base-ten blocks: squares, strips, and circles that represent hundreds, tens, units, tenths, hundredths, or thousandths.

base units: the meter, liter, and gram of the metric system.

brackets: the symbol [ ] which may surround parentheses and help to group numbers in arithmetic.

capacity: the amount of volume that a liquid occupies in a container.

centi: metric prefix for one hundredth.

coefficient: a number that multiplies a variable. In $4x$, for example (which means 4 times x), the coefficient is 4.

commutative property of multiplication: when multiplying numbers, the order doesn't affect the answer: $a \times b = b \times a$.

composite number: a whole number greater than one which isn't a prime number.

congruent: two line segments that have the same length, or two angles that have the same angular measure.

constant: a fixed value. For example, in $x + 7$, the number 7 is the constant.

credit: a transaction that adds to the balance of a checking account.

cube: a three-dimensional solid formed from six congruent perpendicular square sides.

customary units: units of measure common in the United States, like feet, pounds, ounces, quarts, and gallons.

debit: a transaction that reduces the balance of a checking account.

decagon: a polygon with 10 sides.

deci: metric prefix for one tenth.

decimal: a fraction where the denominator is 10, 100, 1000, or other power of ten.

decimal square: a 10×10 grid for visualizing hundredths.

degree: one 360th of a circle.

deka: metric prefix for ten.

denominator: the number at the bottom of a fraction. In $\frac{2}{7}$, for example, the denominator is 7.

dependent variable: a variable that is affected by a change to the independent variable. The dependent variable goes on the vertical axis of a scatter plot.

deposit: money that is added to a bank account.

difference: the result of subtracting two numbers.

dime: a United States coin worth 10 cents.

distributive property: $a \times (b + c) = a \times b + a \times c$.

dividend: the number that is being divided, like the 24 in $24 \div 3 = 8$.

divisor: the number divided by, like the 3 in $24 \div 3 = 8$. The dividend divided by the divisor equals the quotient:

$$\text{dividend} \div \text{divisor} = \text{quotient}$$

dollar: a United States bill worth 100 cents.

dot plot: a graph where each data point is a dot and the dots are stacked in columns on a number line.

equiangular: a polygon for which all of the angles have the same angular measure. All of the angles are congruent.

equilateral: a polygon for which all of the sides have the same length. All of the edges are congruent.

expanded form: when a number is written as a sum based on its digits, such as $642.7 = 600 + 40 + 2 + \frac{7}{10}$.

factor: a number that is being multiplied. For example, in $4 \times 7 = 28$, the numbers 4 and 7 are called factors.

frequency: the number of times that a result occurs in a data set.

GCF: the greatest common factor. For example, the GCF of 16 and 24 equals 8 since $16 = 2 \times 8$ and $24 = 3 \times 8$.

gross income: the total amount earned (before subtracting taxes or making other deductions).

hecto: metric prefix for one hundred.

height: a distance across a triangle that is perpendicular to its base.

heptagon: a polygon with 7 sides.

hexagon: a polygon with 6 sides.

horizontal: an axis on a scatter plot that runs across from left to right. This serves as the x-axis.

identity property of multiplication: multiplying by one has no effect: $1 \times a = a$.

improper fraction: a fraction for which the numerator is larger than the denominator, such as $\frac{5}{4}$.

income tax: money that a person pays to the government based on the person's earnings.

increment: a fixed amount by which a quantity is changed, like the spacing between tick marks on a number line.

independent variable: a variable that either changes on its own or which can be controlled. The independent variable goes on the horizontal axis of a scatter plot.

integer: whole numbers like 0, 1, 2, 3, 4, etc., and negative values like −1, −2, −3, −4, etc. Some books and instructors distinguish between integers and whole numbers, the difference being that integers include negative values.

intercept: the point where a line or curve crosses an axis.

intersect: when two lines or curves cross.

inverse operation: the operation that is the opposite. For example, subtraction is the inverse of addition and division is the inverse of multiplication.

inverse property of multiplication: division is the opposite of multiplication: $a \div a = a \times \frac{1}{a} = \frac{a}{a} = 1$.

inverse proportion: a relationship where an increase in one quantity causes a decrease in the other quantity.

irregular: a polygon that either isn't equilateral or isn't equiangular.

isosceles: a triangle with at least two congruent edges.

kilo: metric prefix for one thousand.

lb: the abbreviation for pound (a customary unit of weight).

LCD: lowest common denominator. Given two fractions, the LCD is the LCM of the two denominators.

LCM: least common multiple. For example, 12 is the LCM of 4 and 6 since $4 \times 3 = 12$ and $6 \times 2 = 12$.

leaf: the units digit of a two-digit number on a stem-and-leaf plot. In 24, for example, 2 is the stem and 4 is the leaf.

metric units: units of measurement where a prefix (such as centi or kilo) is added to a base unit (meter, liter, or gram). For example, the kg combines kilo and gram.

milli: metric prefix for one thousandth.

mixed number: a number that includes an integer plus a fraction, such as $4\frac{2}{5}$ (which equals four plus two-fifths).

multiple: a whole number that can be made by multiplying two smaller whole numbers together. For example, 18 is a multiple of 6 because $6 \times 3 = 18$.

natural number: positive numbers that are whole, like 1, 2, 3, 4, etc. Some books and instructors distinguish between natural numbers and whole numbers, the difference being that natural numbers don't include zero.

net income: the amount earned after subtracting taxes (and after making any other deductions).

nickel: a United States coin worth 5 cents.

nonagon: a polygon with 9 sides.

number line: a horizontal axis with evenly spaced tick marks that helps to visualize numbers.

numerator: the number at the top of a fraction. In $\frac{2}{7}$, for example, the numerator is 2.

obtuse: an angle that is greater than 90°. An obtuse triangle contains one obtuse angle.

octagon: a polygon with 8 sides.

odometer: an instrument displayed on the dashboard of an automobile which measures the distance traveled.

operator: a symbol that represents a mathematical process, such as the times symbol or plus sign.

ordered pair: a pair of numbers in the form (x, y) locating a point on the coordinate plane.

origin: the point $(0, 0)$ on the coordinate plane where the x- and y-axes intersect.

parallel: two lines that are equally spaced apart so as not to intersect (even if they were extended indefinitely).

parallelogram: a quadrilateral with two pairs of parallel edges.

parentheses: the symbols ( ) which help to group numbers in arithmetic.

partial quotients: a method of division where easy multiples of the divisor are subtracted until reaching zero.

payroll tax: money that an employer withholds from a paycheck to pay the government.

penny: a United States coin worth 1 cent.

pentagon: a polygon with 5 sides.

perimeter: the sum of the lengths of the edges of a polygon.

perpendicular: at right angles to one another.

place value: the value of a digit's position in a number. For example, in 1.73, the 3 is in the hundredths place.

polygon: a closed geometric figure bounded by line segments.

power of ten: 10, 100, 1000, etc. or 0.1, 0.01, 0.001, etc. These numbers can be made by multiplying tens (like $10 \times 10 \times 10 = 1000$) or by dividing one by tens (like $1 \div 10 = 0.1$).

prime number: a whole number greater than 1 that is evenly divisible only by itself and one, such as 2, 3, 5, 7, and 11.

prism: a solid geometric figure with ends that are parallel congruent polygons and with sides that are parallelograms.

product: the result of multiplying two numbers.

property tax: money that is paid to the government (usually once per year) for owning property.

proportional: a relationship where a change in one quantity causes a change in another quantity. In a direct proportion, an increase in one quantity causes a corresponding increase in another quantity.

quadrilateral: a polygon with 4 sides.

quarter: a United States coin worth 25 cents.

quotient: the result of dividing two numbers.

range: the difference between the greatest value and the least value in a set of a data. It provides a measure of the full spread of the data.

rectangle: a parallelogram with four 90° interior angles. It is equiangular, but isn't equilateral and isn't regular.

rectangular prism: a three-dimensional solid bounded by six perpendicular rectangles. It is a rectangular box.

reduced fraction: the simplest form of a fraction, where the numerator and denominator do not share a GCF.

regroup: make groups of tens (or hundreds, etc.), like when we subtract $40 - 17 = 23$ by rewriting 40 as 3 tens and 10 ones.

regular: a polygon that is both equilateral and equiangular.

remainder: the amount left over when a number is divided by another number. For example, $17 \div 5$ equals three with a remainder of two since $3 \times 5 = 15$ and $17 - 15 = 2$.

renaming: rewriting a mixed number by borrowing one from the whole number in order to subtract a fraction, like when we use $1\frac{1}{6} = \frac{7}{6}$ to rewrite $3\frac{1}{6}$ as $2\frac{7}{6}$ in order to subtract $\frac{4}{5}$.

rhombus: a parallelogram with four congruent edges. It is equilateral, but it isn't equiangular and it isn't regular.

right: a 90° angle. A right triangle includes one right angle. Perpendicular lines intersect at right angles.

rise: the vertical height between two points when finding the slope of a line.

ruler: a thin strip of plastic, wood, or metal scaled in inches or centimeters used to measure distances shorter than a foot.

run: the horizontal base between two points when finding the slope of a line.

sales tax: money paid when a customer makes a purchase.

scalene: a triangle with no congruent edges.

scatter plot: a graph of ordered pairs that helps to see the relationship between two quantities.

slope: a measure of the steepness of a line determined by dividing the rise by the run.

solve: determine the value of the variable by following a procedure (like isolating the unknown).

spread: an indication of how much the values in a data set vary, such as the range.

square: a quadrilateral that has 90° angles and also has 4 congruent edges. A square is a regular quadrilateral, as it is both equilateral and equiangular.

stem: the tens digit of a two-digit number on a stem-and-leaf plot. In 24, for example, 2 is the stem and 4 is the leaf.

stem-and-leaf plot: a plot that groups data values by their tens digits (called stems).

tablespoon: a measuring spoon equivalent to 3 teaspoons. One fluid ounce equates to 2 tablespoons.

tax: money collected by the government.

teaspoon: a measuring spoon that is three times smaller than a tablespoon. One fluid ounce equates to 6 teaspoons.

term: expressions separated by plus, minus, or equal signs in an algebraic expression or equation. For example, $3x$, 8, and $5x$ are each terms in $3x + 8 = 5x$.

times symbol: the cross ($\times$), like the symbol in $4 \times 3 = 12$.

trailing zeroes: zeroes that come at the end of a number and which follow a decimal point, such as 1.300.

transaction: a process which results in a credit or debit to a bank account.

trapezoid: a quadrilateral with one pair of parallel edges.

triangle: a polygon with 3 sides.

trundle wheel: a measuring device with a wheel that rolls along the ground, which measures distance.

unit: 1) a standard value for making measurements, such as a meter, foot, or second. 2) the ones digit of a number.

unit conversion: a mathematical process that expresses a measurement as an equivalent value in different units, like the process of expressing 4 feet as 48 inches.

unit cube: a cube with edges that are one unit long.

unit square: a square with edges that are one unit long.

variable: an unknown quantity like x or y.

vertex: a point where two lines intersect.

vertical: an axis on a scatter plot that is straight upward. This serves as the y-axis.

volume: the amount of space that an object occupies.

whole number: numbers that are whole, like 0, 1, 2, 3, 4, etc. Some books and instructors distinguish between whole numbers and integers, the difference being that integers include negative values.

withdrawal: money that is taken out of a bank account.

***y*-intercept:** the point where a line crosses the y-axis.

INDEX

A

B

C

D

E

F

G

H

I

K

L

M

N

O

P

R

S

T

U

V

WAS THIS BOOK HELPFUL?

A great deal of effort and thought was put into this book, such as:

- Which topics to cover to meet the needs of students of diverse levels and abilities. Hopefully, the exercises varied from easy to challenging.
- Careful selection of problems for their instructional value.
- Including the answer to every problem, along with many solutions and explanations.
- Examples to help illustrate how to solve the problems.

If you appreciate the effort that went into making this book possible, there is a simple way that you could show it:

Please take a moment to post an honest review.

For example, you can review this book at Amazon.com or Goodreads.

Even a short review can be helpful and will be much appreciated. If you're not sure what to write, following are a few ideas, though it's best to describe what's important to you.

- Are you satisfied with the topics that were covered?
- Did you enjoy the selection of problems?
- Were you able to understand the examples and answer key?
- How much did you learn from reading and using this workbook?
- Would you recommend this book to others? If so, why?

Do you believe that you found a mistake? Please email the author, Chris McMullen, at greekphysics@yahoo.com to ask about it. One of two things will happen:

- You might discover that it wasn't a mistake after all and learn why.
- You might be right, in which case the author will be grateful and future readers will benefit from the correction. Everyone is human.

ABOUT THE AUTHOR

Dr. Chris McMullen has over 20 years of experience teaching university physics in California, Oklahoma, Pennsylvania, and Louisiana. Dr. McMullen is also an author of math and science workbooks. Whether in the classroom or as a writer, Dr. McMullen loves sharing knowledge and the art of motivating and engaging students.

The author earned his Ph.D. in phenomenological high-energy physics (particle physics) from Oklahoma State University in 2002. Originally from California, Chris McMullen earned his Master's degree from California State University, Northridge, where his thesis was in the field of electron spin resonance.

As a physics teacher, Dr. McMullen observed that many students lack fluency in fundamental math skills. In an effort to help students of all ages and levels master basic math skills, he published a series of math workbooks on arithmetic, fractions, long division, percentages, algebra, geometry, trigonometry, logarithms, and calculus entitled *Improve Your Math Fluency*. Dr. McMullen has also published a variety of science books, including astronomy, chemistry, and physics workbooks.

Author, Chris McMullen, Ph.D.

GRADE 6 MATH WORKBOOK

with ANSWERS

order of operations
prime factorization
fractions/decimals
geometric figures
financial math

$$\frac{8}{?} = \frac{48}{72}$$

ratios/proportions
prealgebra skills
data analysis
histograms
exponents

$$(3+9)^2 \div 8 - 5 \times 3$$

Chris McMullen, Ph.D.

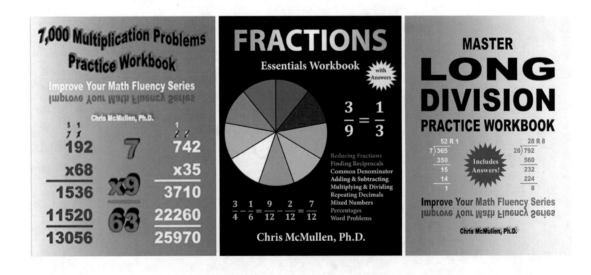

MATH

This series of math workbooks is geared toward practicing essential math skills:

- Fractions, decimals, and percentages
- Long division
- Prealgebra
- Algebra
- Geometry
- Trigonometry
- Logarithms
- Calculus
- Multiplication and division
- Addition and subtraction

www.improveyourmathfluency.com

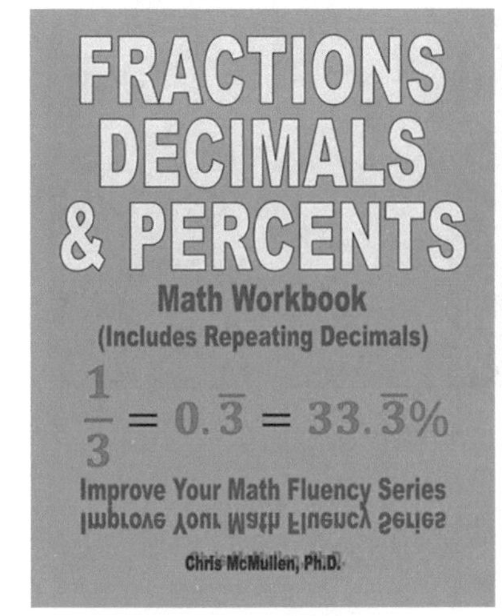

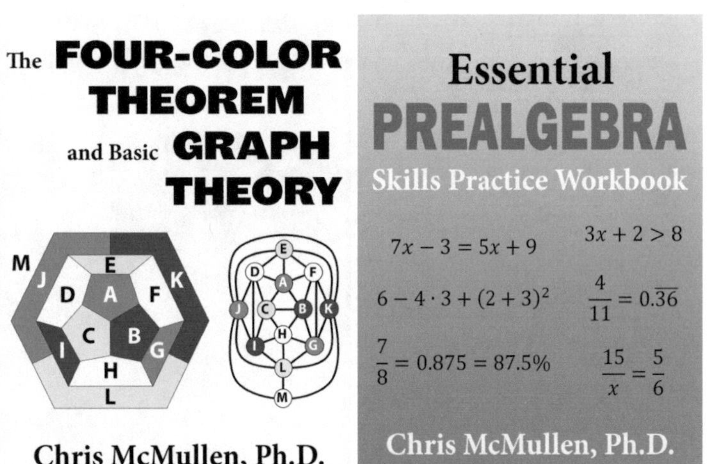

PUZZLES

The author of this book, Chris McMullen, enjoys solving puzzles. His favorite puzzle is Kakuro (kind of like a cross between crossword puzzles and Sudoku). He once taught a three-week summer course on puzzles. If you enjoy mathematical pattern puzzles, you might appreciate:

300+ Mathematical Pattern Puzzles

Number Pattern Recognition & Reasoning

- Pattern recognition
- Visual discrimination
- Analytical skills
- Logic and reasoning
- Analogies
- Mathematics

SCIENCE

Dr. McMullen has published a variety of **science** books, including:

- Basic astronomy concepts
- Basic chemistry concepts
- Balancing chemical reactions
- Calculus-based physics textbooks
- Calculus-based physics workbooks
- Calculus-based physics examples
- Trig-based physics workbooks
- Trig-based physics examples
- Creative physics problems
- Modern physics

www.monkeyphysicsblog.wordpress.com

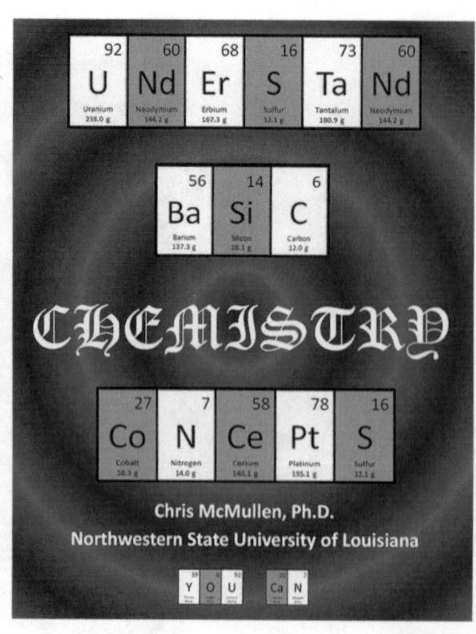

Made in the USA
Thornton, CO
05/27/24 08:18:33

c5f592ea-0640-4065-81f0-5280b7247065R01